Payment for Ecosystem Services

ECOLOGICAL ECONOMICS AND HUMAN WELL-BEING

Series Editors:

Joan Martínez-Alier, Professor of Economics and Economic History at Universitat Autonoma de Barcelona. He is also the founding member and former President, International Society for Ecological Economics (ISEE).

Pushpam Kumar, Environmental Economist, Institute for Sustainable Water, Integrated Management and Ecosystem Research (SWIMMER), and Department of Geography, University of Liverpool, UK.

Technical Editor: Virginia Hooper

This Series offers the best and most recent works in the transdisciplinary field of Ecological Economics, whose focus is the study of the conflicts between economic growth and environmental sustainability.

Books in the Series:

Payment for Ecosystem Services

Pushpam Kumar
and
Roldan Muradian

OXFORD
UNIVERSITY PRESS

OXFORD
UNIVERSITY PRESS

YMCA Library Building, Jai Singh Road, New Delhi 110 001

Oxford University Press is a department of the University of Oxford.
It furthers the University's objective of excellence in research, scholarship,
and education by publishing worldwide in

Oxford   New York
Auckland   Cape Town   Dar es Salaam   Hong Kong   Karachi
Kuala Lumpur   Madrid   Melbourne   Mexico City   Nairobi
New Delhi   Shanghai   Taipei   Toronto

With offices in
Argentina   Austria   Brazil   Chile   Czech Republic   France   Greece
Guatemala   Hungary   Italy   Japan   Poland   Portugal   Singapore
South Korea   Switzerland   Thailand   Turkey   Ukraine   Vietnam

Oxford is a registered trademark of Oxford University Press
in the UK and in certain other countries.

Published in India
by Oxford University Press, New Delhi

ISBN-13: 978-0-19-569874-9
ISBN-10: 0-19-569874-6

Typeset in AGaramond 11/12.5
by Eleven Arts, Keshav Puram, Delhi 110 035
Printed in India by Pauls Press, New Delhi 110 020
Published by Oxford University Press
YMCA Library Building, Jai Singh Road, New Delhi 110 001

Contents

Tables, Figures, Maps, and Boxes

FIGURES

MAPS

Boxes

Abbreviations

AIC	Akaike Information Criterion
BMC	Bhopal Municipal Corporation
BRMP	Buccoo Reaf Marine Park
CBD	Convention on Biological Diversity
CER	Certified Emission Reduction
CGR	Crop Genetic Resource
CI	Confidence Interval
CRP	Conservation Research Programme
CS	Consumer Surplus
CV	Contingent Valuation Method
EDC	Eco-development Committees
EES	Ecological and Environmental Services
EQIP	Environmental Quality Improvement Programme
FECF	Forest Ecosystem Compensation Fund
HH	Household
HYV	High Yielding Variety
IAD	Institutional Analysis and Development
IIA	Independence from Irrelevant Alternatives
IID	independently and identically distributed
IIED	International Institute for Environment and Development
JBIC	Japan Bank for International Cooperation
LCA	Lake Conservation Authority
LCM	Latent Class Model
LNR	Lashihai Nature Reserve
LOWK	Lower Kolans Watershed
MAB	Municipal Agricultural Bureau
MBI	market based instruments
MNL	multinomial logit
MPCE	monthly per capita expenditure
MTB	Municipal Tourism Bureau
MWTP	marginal willingness to pay
NABARD	National Bank for Agriculture and Rural Development

NGO	non-governmental organization
NOAA	National Oceanic and Atmospheric Administration
NSS	negotiation support systems
OBC	other backward caste
OECD	Organisation for Economic Co-operation and Development
PES	payment for environmental ecosystem services
PPT	Pro-Poor Tourism
PWS	Payment for Watershed Services
RP	revealed preference
RSPB	Royal Society for Protection of Birds
SBI	State Bank of India
SC	Scheduled Caste
SD	Standard deviation from means
SLCP	Sloping Land Conversion Programme
SP	stated preference
ST	Scheduled Tribes
STR	Sundarban Tiger Reserve
SURE	Seemingly Unrelated Regression Equations
TC	travel cost
TCC	tourism carrying capacity
TCM	travel cost method
UPK	Upper Kolans Watershed
USP	Unique selling proposition
WES	Watershed Environmental Services
WII	Winrock International India
WTAC	willingness to accept compensation
WTP	willingness to pay

Foreword

At the conference of the International Society for Ecological Economics (ISEE), in New Delhi in December 2006 where the majority of these papers were first presented, 'economic valuation and payment for environmental services' was most analysed. In the current series on Ecological Economics and Human Well-Being, a volume on these issues is indeed timely.

However, a few days before the conference, a short article in *Nature* (7 September 2006, no. 443) by Douglas Mc Cauley of Stanford University had dismissed payment for environmental services as a fashion. Douglas argued that nature provided ecological and aesthetic values. To these, one would add the direct livelihood value of poor people. Such values were forgotten under the trend in conservation science that saw 'ecosystem services' as economic benefits provided by natural ecosystems. According to him, it was assumed that if scientists could identify ecosystem services, quantify their economic value, and ultimately bring conservation into the market, the owners of the environmental services would have an incentive to preserve and increase them, and the public decision makers would recognize the folly of environmental destruction and work to safeguard nature.

The reader might agree or disagree with this simplified version of a more complex issue. Precisely, the following chapters analyse a number of concrete instances in which economic valuation and, sometimes, payment for environment services have taken place. Distinctive of this volume are the many examples from the South, such as India, Nepal, China, and countries in Africa and Latin America. Also included are case studies from Australia and other rich countries. The ecosystems and ecosystem services that are considered here range from wetlands to mountain areas. These include pollination services, preservation of agricultural varieties of rice, provision of clean water, etc. The volume discusses the disputes on property rights on the environment that are prerequisite to any payment to happen, and whether PES schemes lead or not lead to a less unequal income distribution.

The volume editors, as also many of the authors in the volume are well known experts on valuation and payment for environmental services. Roldan Muradian has written mainly on Latin America. He has a background as a biologist in Venezuela, and is now a well known ecological economist based in The Netherlands. Pushpam Kumar is an economist who has been closely associated with the ISEE and the Indian Society for Ecological Economics, and has taught ecological and development economics in UK and India. He has published many papers on the economic valuation of ecosystem services of wetlands, mangroves, and forests in India and Europe, insisting on the notion of 'the GDP of the poor', referring to the fact that natural systems provide directly, outside the market, a large part of the livelihood of poor people.

October 2008   Joan Martínez-Alier
                                                 Pushpam Kumar

1

Payment for Ecosystem Services and Valuation
Challenges and Research Gaps

PUSHPAM KUMAR AND ROLDAN MURADIAN

INTRODUCTION

Management of environmental services is an urgent issue, given that their provision by natural ecosystems worldwide is declining due to human interventions. For example, more land has been converted to cropland since 1945 than that in the 18th and 19th centuries taken together; 25 per cent of the world's coral reefs have been badly degraded or destroyed in the last decades; 35 per cent of mangrove area has been lost during this time; the amount of water in reservoirs has quadrupled since 1960 and withdrawals from rivers and lakes have doubled since 1960 (MA 2005). Valuation of ecosystem services and the subsequent design of compensation schemes have been proposed as effective tools that may considerably reduce the cost of such management. The ultimate purpose of valuation and payments for ecosystem services is to estimate and compensate the costs incurred by providers and to highlight economic efficiency considerations for decision makers in the concerned location. Valuation provides insight into the losses (or gains), across different stakeholders, arising from perturbances in ecosystems and their services. Furthermore, payments for environmental (or ecosystem) services (PES) have been proposed as redistributive mechanisms between different social groups, and the whole issue of benefits emanating from different ecosystems could also be framed in the context of inequality concerns in rural–urban dynamics.

The rural sector is facing daunting challenges worldwide. For the first time in human history, people living in urban areas will soon surpass the rural population. Furthermore, in most countries, the relative economic importance of the rural vis-à-vis the overall economy (measured, for instance, as the share of rural gross domestic product or GDP as percentage of total GDP) has been declining steadily for many decades. Deterioration in the terms of trade of agriculture in relation to other economic sectors seems to be a structural feature of the capitalist way of development. In addition, due to a variety of reasons, the current process of economic globalization favours economies of scale in the agri-business sector, which is accelerating the market exclusion of small farm holders in developing countries. In addition, heavy subsidies to farmers in developed countries are considerably hindering market opportunities to farmers from developing countries—thus contributing towards increase in global inequality—and are absorbing large amounts of resources that could be invested either in more dynamic economic activities, in closing the global development gap, or in research for solving acute problems, such as global warming. The rural crisis in developing countries has considerable implications for North–South relations, since it is a major driving force of, for example, large scale South–North migration (a labour flow that is, in part, devoted to underpaid agricultural work in the North). Equally, the rural crisis tends to contribute to worsen living conditions in poor urban areas in the developing world. This general context, entailing serious constraints on rural development worldwide, calls for new approaches to redefine the economic functions of rural areas, in both developed and developing countries. Some authors have proposed that innovative schemes for compensating the provision of environmental services may become the basis of a new urban–rural compact, which would revolutionize economic relations between these sectors (Smith 2006; Wolcott 2006; Gutman 2007). Is the provision of environmental services able to mobilize enough resources to improve the living conditions of rural populations, particularly in developing countries? Can a new agenda for compensating for these services trigger a radical change in the way subsidies are allocated to rural dwellers in developed countries? Which methods should be used to quantify the supply of environmental services and the trade-offs landholders face in relation to alternative land uses? This book aims to help find some answers to these questions.

This introductory chapter is structured as follows. First, we carry out a brief review of the literature on PES and the contributions included as chapters in the book. Second, we characterize the major challenges of PES schemes for becoming a driving force of rural development worldwide. Last, we systematize what, according to us, are the main research gaps and try to set a possible research agenda.

Ecosystems and biodiversity provide a wide range of services through bio-geo-chemical processes that are critical for the sustainability of human societies. Ecosystems provide services that sustain, strengthen, and enrich various constituents of human well-being. The unique feature of most of the services emanating from ecosystems is that—although acknowledged by people—they are unaccounted, un-priced, and therefore remain outside the domain of the market. In conventional parlance, such problems are treated as externalities where market fails, and decision makers try to correct the market failure by creating market-like situations. To do so, they need to estimate the value of services through various valuation techniques based on stated or revealed preferences of the people. The valuation of services related to the provision of natural resources is relatively easier to quantify, as such services are normally closer to the market domain. On the other hand, regulating, supporting, and cultural services are difficult to capture and are known for a high uncertainty with regards to their biophysical production functions (Kumar and Kumar 2007). However, that definitely does not imply that these services should not be accounted and evaluated monetarily, as they play a critical role in the sustainability of economic processes.

In order to provide true and meaningful indicators of ecosystem values, economic valuation should account for the state of ecosystems and ecosystem functions. For example, special attention should be given to non-linear and thresholds effects (Holling 2001). Standard economic tools based on marginal analytic methods are limited to situations when ecosystems are relatively intact and functioning in normal bounds far away from any bifurcation (Limburg et al. 2002). This is of particular significance to developing countries, wherein significant trade-offs exist between conservation and economic development, and decisions often favour the latter. Therefore, decisions made based on 'snapshot' ecosystem values can provide false policy directives.

Another critical issue deals with summing up individual values to arrive at aggregate values, viz. 'societal values'. Ecosystem goods and services, by definition, are public in nature, meaning thereby that several benefits accrue to society as a whole, apart from the benefits provided to the individuals (Daily 1997; Wilson and Howarth 2002). The theoretical fundamentals of economic valuation methodologies rest on the axiomatic approaches of individual preferences and individual utility maximization, which do not always match the public good characteristic of ecosystem services. Valuation methodologies such as contingent valuation utilize individual preferences as a basis for deriving values used for resource allocation of goods that are largely public in character. A considerable body of recent literature, therefore, favours the adoption of a discourse-based valuation (Wilson and Howarth 2002). The primary focus of these approaches is to utilize a discourse-based valuation approach to come up with a consensus on societal value of scarcity indicator, derived through a participatory process, to be used for allocation of ecological services largely falling into the public domain. Due to the rising and fashionable use of the terms 'environmental services' and 'payments for environmental services', some authors have tried to develop rigorous definitions (Rosa et al. 2004; Wunder 2005; Boyd and Banzhaf 2007; Ravnborg et al. 2007; Pagiola and Platais, 2007). Despite the lack of consensus about the meaning—which is unlikely to be reached—PES in general refers to economic transfers aiming to compensate agents (normally landholders) for the provision of positive environmental externalities, which in principle should entail opportunity costs for providers. The implementation of PES has been rather recent. Nonetheless, some schemes have been already in place for enough time in order to draw some lessons. The four main services for which PES have been created so far are carbon sequestration, biodiversity conservation, hydrological services, and landscape beauty. To date, research on PES has been focused on efficiency considerations (Landell-Mills 2002; Wunscher et al. 2006; Wunder 2007), the effects of tenure security and economic valuation (Ahlheim and Neef, 2006), the impacts on poverty reduction (Grieg-Gran et al. 2005; Pagiola et al. 2005; Tschakert 2005), and institutional issues (Pascual and Perrings 2007). The contents of the chapters composing this book are very diverse, and contribute to most of the issues mentioned above. However, they may be divided into two main thematic blocks: those dealing with methodological applications of valuation techniques to environmental services and those addressing conceptual issues related to PES design, functioning, and assessment. In principle, PES can be used in diverse contexts and for most of the services. However, the ability to meet the expected goals may vary considerably, depending on a number of factors. For example, Mayrand and Paquin (2004) evaluated more than 300 instances of PES and concluded that these mechanisms have a larger probability to be successful at a local scale, rather than at the global level. In addition, the success of valuation and PES critically depends upon the clarity with which a service is perceived by the beneficiaries. Furthemore, the transaction costs of the entire process must not exceed the benefits and finally, PES needs social trust, effective governance, and explicit institutional arrangements. All these aspects are tackled to a large or lesser extent in this book. The following paragraphs summarize the main issues addressed by each of the chapters composing this volume.

The second chapter in this volume by Wendy Proctor, Thomas Köllner, and Anna Lukasiewicz sets up the methodology for assessing equity and fairness issues in market-based instruments for improving environmental services, and then exemplifies the method through a case study that involves the use of an economic tool to combat salinity recharge in a farming area of Victoria, Australia. The aim of the chapter is not to provide a definitive study on equity issues in market-based instruments for environmental protection, but to raise issues concerning equity that should be incorporated into these schemes. The third chapter by Stefanie Engel and Charles Palmer develops a model for analysing the design of PES, when the actors taking the land-use decisions have only weak property rights over the land, and when other actors (such as firms) interested in commercial resource exploitation are present—a situation that is common in developing countries. In such a context, designing PES is a complex task, and common intuition may be quite misleading. In particular, we will show that—if not designed carefully—PES may lead to resource owners negotiating better exploitation agreements with commercial actors (for example, better logging deals) rather than environmental benefits. Moreover, the authors point out that designing PES to achieve maximum environmental benefits generally does not involve targeting of the poorest communities. This poverty–environment trade-off and the potential of property rights reforms as a complementary policy option are also discussed. The fourth chapter by Sunandan Tiwari and Jaime Amezaga examines negotiation support systems for environmental services derived from watersheds. The authors present an analytical framework that draws on institutional assessment. The framework assists in analysing local and external factors and actors in order to identify key incentives, both within and outside the local community, for improving watershed environmental services. It allows for the analysis of involved institutions in a manner that provides insights into the nature of the negotiations utilized as well as the sustainability of the mechanism. This analytical tool is being tested on an ongoing action-learning project aimed at developing incentive-based mechanisms for watershed environmental services in the Bhoj Wetlands of Madhya Pradesh, India.

The fifth chapter by R.A. Hope, M. Borgoyary, and C. Agarwal explores the issue of economic incentives for small-scale farmers in order to adopt organic farm management as an approach to mitigate off-site water quality impacts at the Bhoj Wetlands, India. The authors found that a choice experiment is a suitable approach to estimate farmer preferences to inform policy design, and that incentives are key factors for the adoption of organic practices, subject to farm location, farmer profile, and current farming practices. Their results suggest that if farmers can access premium organic markets, land certification may provide a self-enforcing institutional arrangement for land management change. This would constitute a compensation mechanism for PES.

The sixth chapter by Muo J. Kasina, John Mburu, and Karin Holm-Mueller uses the contingent valuation approach to assess local communities' willingness to pay (WTP) for enhanced pollination services for farmlands surrounding the Kakamega forest, a biodiversity hotspot in Kenya. The study addresses the factors that influence the magnitude of farmers' WTP. A number of limitations and implications of using particular payment vehicles are discussed, focusing especially on the conversion of in-kind-payments to monetary values in developing countries such as Kenya. The seventh chapter by C. Giupponi, A. Goria, A. Markandya and A. Sgobbi analyses the feasibility of using economic compensations for the provision of environmental services in order to address problems related to natural resources management in the Lashihai Nature Reserve in the Yunnan Province of China. The two main problems this study addresses are the estimation of a suitable level of compensation to farmers for the damages inflicted to their yields by protected bird species; and the most suitable design of an incentive mechanism to induce farmers to adopt more environmental-friendly agricultural practices, in order to provide clean water for landscape services to the tourist city of Lijiang. The eighth chapter by Wendy Proctor, Jeffery

Connor, John Ward, and Darla MacDonald describes a comprehensive method to design, test, and then implement a PES framework to combat the environmental consequences of extensive native vegetation clearance in Australia. The scheme, based on a voluntary 'cap and trade' approach, allows farmers to meet recharge obligations by land management actions or by trading credits. Assessment of the scheme so far suggests that an incentive for aggregate group outcome achievement included in the design may have motivated higher enrolment rates than would have otherwise occurred. The ninth chapter by Manuela Gaehwiler, Brian Gross, Thomas Koellner, Wendy Proctor, and David Zilberman investigates the level of marketing performance of selected suppliers of environmental services in Latin America. The results of the study show that marketing plays, as yet, only a minor role in business planning and so does the evaluation of the demanders' needs. Customers' needs and expectations are hardly known and hence the products are not designed accordingly. They conclude that the lack of marketing strategies is likely to be hindering the diffusion of PES and that improved marketing, understood in a comprehensive way, may help the supply side to address the needs of potential buyers and to increase market transactions.

The tenth to thirteenth chapters address some of the nuances of valuation of ecosystem services so pivotal to success for their market.

The tenth chapter by Nesha Beharry and Riccardo Scarpa reports the results of a choice experiment to investigate the WTP for an improvement in coastal water quality for beach recreation in Tobago and illustrates how these values vary by the type of recreator. This study shows the importance of using preference elicitation and estimation methods that account for differences across individual preferences for coastal ecosystem services. The inclusion of individual preferences in multi-attribute valuation studies may help to improve the prioritization of policy initiatives by incorporating the diverse attributes of complex ecological systems. The eleventh chapter by Diwakar Poudel and Fred H. Johnsen analyses the main factors influencing the WTP for conservation of rice landraces in Kaski, Nepal. The study indicates that farmers value crop genetic resources (such as rice landraces) for use, non-use, and existence options and there is considerable support for conservation. Farmers consider community gene banks as a best alternative method because the genetic resources are not only conserved but also utilized and farmers have access and control over the resources in such genetic management schemes. This study contributes to the development of methods for the assessment and design of conservation strategies for agro-biodiversity. The study also suggests that the conservation strategy should be adapted to different genetic resources, based on use values and threats to extinction. The twelfth chapter by Ayele Gelan estimates non-market economic benefits of forest recreation in Great Britain. The study employs the travel cost (TC) method of recreation demand modelling, which is based on revealed preference of users who incur a cost to visit a forest site. This study draws attention to deficiencies in the literature, particularly with regard to econometric model specification. The thirteenth chapter by Indrila Guha and Santadas Ghosh sheds light on the socio-economic impact of tourism in Indian Sundarbans in the village Pakhiralay, which holds a unique mangrove delta forest and a World Heritage Site. The study finds evidence of greater inequality in the village economy resulting from tourism and no significant trickle-down effect. It also shows a significant degree of heterogeneity in villagers' perception about important non-measurable issues relating to tourism. Furthermore, the study does not find conclusive evidence that tourism, on its own, has already reduced the locals' forest dependence; though it might lead to a different outcome once the scale of operation increases.

CHALLENGES OF PES SCHEMES FOR BECOMING A DRIVING FORCE IN RURAL DEVELOPMENT

This section systematizes the main challenges that PES face in becoming a major engine for rural development.

Demand

Hitherto, the economic effects of changes in the environmental services provided by rural dwellers to other sectors of society were not taken into account in their decisions about land use, since cash flows were not in place. PES schemes aim to change this situation by means of creating new markets for the supply of such services and, therefore, depend very much on perceptions about the provision functions, and the actual demand for them by other economic agents. This implies that it is not enough to demonstrate that the land use changes promoted by the PES scheme contribute towards enhancing particular environmental services, but it is extremely crucial to ensure that agents on the demand side are informed, share the perception that the promoted land uses have a positive effect on the supply and have an actual demand for the offered services. Thus, raising awareness among beneficiaries should be a key component of any paradigmatic shift.

A shift from the current paradigm guiding the allocation of agricultural subsidies in developed countries based on productivity, to a new framework relying on the provision of environmental services would considerably improve the environmental performance of these countries, while at the same time contributing towards ameliorating global income distribution by means of increasing demand for agricultural products from developing countries. As stated before, such change of paradigm would require raising awareness, particularly among policy makers and urban dwellers. Since developed countries already allocate considerable resources to redistribute wealth from urban to rural areas, such a change in paradigm would not need considerable additional resources, but simply a change in the criteria used to allocate the same. The situation in most developing countries is quite different, since no major urban–rural wealth redistribution systems are in place. There are structural reasons for this fact. First, in developing countries a considerable share of the total population still lives in rural areas. Second, the surplus of the urban sector is usually not sufficient to sustain the allocation of large subsidies to rural dwellers. Hence, a large-scale compensation of landholders in developing countries is significantly harder. Nonetheless, many of the environmental services provided by rural dwellers in developing countries may have a global demand, particularly in the case of carbon sequestration, biodiversity conservation, and landscape beauty. Global transfers of resources for these services already exist and are likely to increase in the future, as they become even scarce more. Nevertheless, at the time being it seems unlikely that the global demand for the environmental services provided by rural areas in developing countries could trigger the appearance of a new rural development compact based on acknowledging the economic importance of environmental services. Moreover, it is very likely that a shift in the way rural subsidies are allocated in developed countries would be accompanied by an expansion of the agricultural frontier in the developing world, which would have a negative effect on the provision of environmental services. In addition, new trends in global energy demand, such as rising demand for bio-fuel and the booming Chinese appetite for energy and materials, are already steering significant land use changes in developing countries, which are, again, not compatible with the provision of environmental services. In summary, a shift in the current paradigm, steering urban–rural transfers of resources in the developed world, is likely to take place if efforts are devoted to change the mindset of policy makers and citizens. However, at the global level, this change seems considerably less likely given the fact that most developing countries are unable to implement large-scale transfers of economic resources between urban and rural areas.

Moral hazards and perverse effects

PES schemes are based on the Coase Theorem; thus they rely on the proposition that environmental externalities may be internalized independently of the initial allocation of property rights. However, the fact that markets for environmental services do not necessarily adopt the 'polluter pays' principle—which has been the guiding principle of modern environmental policy—may pose some serious moral problems (Salzman 2005). To what extent should rural dwellers responsible for deforestation and water pollution, for example, be 'compensated' for changing their environmentally unfriendly practices? The fact that PES schemes, in order to maximize their environmental additionality, tend to compensate agents who have not adopted better environmental practices on their own makes them morally problematic(even though they may be economically efficient) since agents that have been more environmentally friendly in the past may tend to be excluded from the compensation scheme. In addition, payments for promoting more environmentally friendly land use practices may trigger perverse effects, such as inducing new agents to adopt the undesirable practices, just as a strategy to be compensated. Markets are not value-free endeavours. On the contrary, their boundaries and functioning are determined to a considerable extent by social values, including notions of fairness and legitimacy, and these issues are present in the provision of public environmental goods.

Causal models for relating land uses and the provision of environmental services

One of the most critical issues of PES schemes is to ensure that the land use practices promoted will produce the expected changes in the provision of environmental services. The provision of environmental services normally depends on a variety of ecosystem functions, which often are also affected by human interventions in a way that is not fully predictable. Therefore, decisions about the design of PES schemes often have to be taken in a context characterized by a high degree of uncertainty with regard to the causal links between the land use adopted/promoted and the provision of environmental services. This is particularly relevant in the case of hydrological services, since the relationship between land use and hydrological services (impacts on water availability, regulation, and quality) is normally very complex. PES schemes are often designed and implemented with insufficient information regarding these causal relationships, due to the fact that background studies are usually very costly and time-consuming and implementing agencies have scarce resources. The design of PES schemes based upon false assumptions about causal relationships between land use and the provision of environmental services constitutes a major risk of these kinds of tools (Kosoy et al. 2007).

Additionality

In order to meet efficiency requirements, PES schemes have to ensure that the monetary transfers in reality entail the implementation of practices that otherwise would not be in place. For example, if upstream landholders in a watershed are compensated for conserving the forest, the scheme would produce an additional effect only if in its absence agents would indeed chop down the trees. Although demonstrating additionality is clearly a very hard task, it is crucial if practitioners aim to prove that PES are more efficient than alternative policy tools. This issue is, however, often neglected in PES design.

Monitoring is another topic that deserves special attention by PES designers. Environmental services are often characterized by credence attributes, namely their main features cannot be verified directly by the consumer. Therefore, a third-party verification process is needed. However, conducting of monitoring (verification) for both the practices to be adopted (changes in land uses) and the provision of the environmental services (carbon sequestration, water quality, and biodiversity) tends to be costly, and therefore contributes to increasing transaction costs. In developing countries especially with small-scale schemes, transaction costs play a considerable role in determining the feasibility of PES, since they may substantially increase the amount to be collected from users. In such cases, there is likely to be a trade-off between the financial feasibility—keeping the amount to be charged within the WTP of users—and the ability of the system to ensure that the expected changes in the provision of the service have indeed taken place.

In summary, even though the economic logic of PES is robust and straightforward, a large scale implementation of such tools as a driving force of rural development may face a number of constraints, mainly because of two reasons: (i) in general, users are unable to evaluate by themselves the effects of PES on the provision of environmental services; and (ii) due to structural complexity and lack of information, the relationship between land use practices and the provision of environmental services is often difficult to predict and characterize, and thus holds a high degree of uncertainty. This imparts a critical role to third-party verification and scientific inputs. These factors increase transaction costs, enlarge the role and power of intermediary agents, and may create constraints to demand. The combination of all these factors, including the risks of moral hazard, shed doubts on the possibilities of large-scale applicability of PES schemes, particularly in developing countries.

RESEARCH GAPS

In the remaining part of this introductory chapter we attempt to identify key issues that deserve further and in-depth research in the field of market-based instruments for enhancing the provision of environmental services.

Trade-offs analysis of land-use changes

The trade-offs landholders face when deciding about alternative land uses lie at the core of the economic analysis of PES (Börner et al. 2007). In principle, the amount of compensation should be at least equal to the opportunity cost of the land use practices that are being promoted. However, in practice, particularly in developing countries, opportunity costs are hard to estimate and may vary considerably across time. Furthermore, decisions about the conservation of environmental functions at the farm level are regularly complex, and hence difficult to seize. Social norms, habits, and regulations also play a role in inducing particular practices. In addition, strategies for risk management at the farm level are usually very important in influencing decisions about land uses. Thus, potential earnings from alternative land uses are only one of the elements that landholders take into consideration when deciding on land use changes. Therefore, traditional cost–benefit evaluation tools are not always useful for analysing land use trade-offs, and it is necessary to develop integrative methods that would enable the researcher to understand the local driving forces of land use changes, and the underlying rationale of landholders. In order to be realistic, these methods should assume that landholders have a bounded rationality and a complex decision making function about land use, which apart from land rent considerations includes habits, perceptions of environmental functions, food self-reliance considerations, risk management, as well as social regulations and relations.

Methods for identifying and quantifying environmental services

As stated before, since the provision of environmental services cannot often be directly assessed by the beneficiaries, the role of information flow in these markets becomes crucial. However, most current research methods for identifying and quantifying these services tend to be costly and, resources allocated to background studies in PES design, particularly in developing countries, are normally small. Resources allocated to information gathering and communication cannot be large in PES schemes, because they increase transaction costs and large transaction costs are among the main constraints for the implementation of these tools. It is then very necessary to develop innovative, effective, easy to implement, and less costly methods for assessing environmental services, which should shed outcomes easy to communicate, not only to the academic community but also to the beneficiaries and other agents involved. Special attention should be paid to the role of uncertainty in the provision of environmental services, and the perception of different stakeholders about it. In order to enhance legitimacy under uncertain conditions, it is necessary to emphasize stakeholders' participation and deliberation as key elements of the methods.

Institutional aspects

Institutional aspects have not been sufficiently addressed in the PES literature. However, local institutions shape the functioning of markets for environmental services to a large extent. Most of these markets are unlikely to emerge without external intervention and/or to thrive without an intermediary. Thus, the management and reduction of transaction costs are at the core of PES performance. In addition, control over incomplete and uncertain information confers considerable power to intermediary agents, including the ability to enforce payments (for example, through the water bill in the case of water-related environmental services), even though users are not totally aware of its purpose. Thus, the role of management of transaction costs in the feasibility, functioning, and effectiveness of PES schemes deserves more research attention, as well as the role of reputation and social relations between the agents involved.

The implications of changes in the allocation of (often de facto) property rights is another institutional aspect that needs further research. As stated before, conventions and beliefs—which are not always supported by scientific evidence—have played a significant role in the shift in allocation of property rights over land use that have taken place in PES schemes. Lastly, post-contractual activities (such as monitoring and enforcement of agreements) are critical parts of the system, since only through them can the efficiency gains of the scheme be assessed. However, it is common to find, particularly in developing countries, schemes with poorly implemented post-contractual activities. Understanding the reasons for this and the implications for the long-run sustainability of the systems is a topic that should be further investigated.

REFERENCES

Ahlheim, M. and A. Neef (2006), 'Payments for Environmental Services, Tenure Security, and Environmental Valuation: Concepts and Policies Towards a Better Environment', *Quarterly Journal of International Agriculture*, Vol. 45, No. 4, pp. 303–17.

Börner, J., Mendoza, A., and S. Vostic (2007), 'Ecosystem Services, Agriculture, and Rural Poverty in the Eastern Brazilian Amazon: Interrelationships and Policy Prescriptions', *Ecological Economics*, Vol. 64, No. 2, pp. 356–73.

Boyd, J. and S. Banzhaf (2007), 'What are Ecosystem Services? The Need for Standardized Environmental Accounting Units', *Ecological Economics*, Vol. 63, pp. 616–26.

Daily G.C. (ed.) (1997), *Nature's Services: Societal Dependence on Natural Ecosystems*. Island Press, Washington D.C.

Grieg-Gran, M., I. Porras, and S. Wunder (2005), 'How can Market Mechanisms for Forest Environmental Services help the Poor? Preliminary Lessons from Latin America', *World Development*, Vol. 33, No. 2, pp. 237–53.

Gutman, P. (2007), 'Ecosystem Services: Foundations for a New Rural–Urban Compact', *Ecological Economics*, Vol. 62, pp. 383–7.

Holling, C.S. (2001), 'Understanding the Complexity of Economic, Ecological, and Social Systems', *Ecosystems*, Vol. 4, pp. 390–405.

Kosoy, N., M. Martinez-Tuna, R. Muradian, and J. Martinez-Alier (2007), 'Payments for Environmental Services in Watersheds: Insights from a Comparative Study of Three Cases in Central America', *Ecological Economics*, Vol. 6, pp. 446–55.

Kumar, M. and P. Kumar (2007), 'Valuation of Ecosystem Services: A Psycho and Cultural Perspective', *Ecological Economics*, Vol. 64, No. 4, pp. 808–19.

Landell-Mills, N. (2002), 'Developing Markets for Forest Environmental Services: An Opportunity for Promoting Equity while Securing Efficiency?', *Philosophical Transaction of the Royal Society*, Vol. 360, pp. 1817–25.

Limburg, K.E., R.V. O'Neill, R. Costanza, and S. Farber (2002), 'Complex Systems and Valuation', *Ecological Economics*, Vol. 41, pp. 409–20.

Mayrand, K. and M. Paquin (2004), *Payment for Environmental Services: A Survey and Assessment of Current Schemes*, Commission for Environmental Cooperation of North America, Montreal.

Pagiola, S., A. Arcenas, and G. Plantais (2005), 'Can Payments for Environmental Services Help Reduce Poverty? An Exploration of the Issues and the Evidence to Date from Latin America', *World Development*, Vol. 33 No. 2, pp. 237–53.

Pagiola, S. and G. Platais (2007), *Payments for Environmental Services from Theory to Practice*, World Bank, Washington D.C.

Pascual, U. and C. Perrings (2007), 'Developing Incentives and Economic Mechanisms for in situ Biodiversity Conservation in Agricultural Landscapes', *Agriculture, Ecosystems and Environment*, Vol. 121, pp. 256–68.

Ravnborg, H., M. Damsgaard, and K. Raben (2007), 'Payments for Ecosystem Services, Issues and Opportunities for Development Assistance', Danish Institute for International Studies, DIIS Report 2007, Vol. 6, Copenhagen.

Rosa, H., S. Kandel, and L. Dimas (2004), 'Compensation for Environmental Services and Rural Communities: Lessons from the Americas', *International Forestry Review*, Vol. 6, No. 2, pp. 187–94.

Salzman, J. (2005), 'The Promise and Perils of Payments for Ecosystem Services', *International Journal of Innovation and Sustainable Development*, Vol. 1, Nos 1 and 2, pp. 5–20.

Smith, K. (2006), 'Public Payments for Environmental Services from Agriculture: Precedents and Possibilities', *America Journal of Agricultural Economics*, Vol. 88, No. 5, pp. 1167–73.

Tschakert, P. (2005), 'Environmental Services and Poverty Reduction: Options for Smallholders in the Sahel', *Agricultural Systems*, Vol. 94, pp. 75–86.

Wolcott, R. (2006), 'Prospects for Ecosystem Services in the Future Agricultural Economy: Reflections of a Policy Hand', *America Journal of Agricultural Economics*, Vol. 88, No. 5, pp. 1181–3.

Wunder, S. (2005), 'Payment for Environmental Services: Some Nuts and Bolts', CIFOR Occasional Paper, Vol 42, CIFOR, Bogor.

———. (2007), 'The Efficiency of Payments for Environmental Services in Tropical Deforestation', *Conservation Biology*, Vol. 21, No. 1, pp. 48–58.

Wunscher, T., S. Engel, and S. Wunder (2006), 'Payments for Environmental Services in Costa Rica: Increasing Efficiency through Spatial Differentiation', *Quarterly Journal of International Agriculture*, Vol. 45, No. 4, pp. 319–37.

Wilson, M.A. and R.B. Howarth (2002), 'Discourse-based Valuation of Ecosystem Services: Establishing Fair Outcomes through Group Deliberation', *Ecological Economics*, Vol. 41, pp. 431–43.

2

Equity Considerations and Payments for Ecosystem Services

WENDY PROCTOR, THOMAS KÖLLNER, AND
ANNA LUKASIEWICZ

INTRODUCTION

Market instruments to enhance or protect ecosystem services or the services provided to humans by nature are increasingly being proposed by environmental economists and policy makers alike as an efficient and cost-effective solution. Such solutions attempt to bestow property rights on services such as carbon sequestration or biodiversity in an attempt to place some monetary value on these previously unvalued services and to encourage individuals to realize their worth with a view to enhancing and saving them. Such solutions are based on efficiency-related outcomes being the overriding goals. Increasingly however, equity issues are being raised as possible deterrents to individuals taking part in such PES or market-based instrument (MBI) schemes and as such these schemes are failing to meet the necessary sustainability objectives of natural resource management. For example, a recent study by Landell-Mills and Porras (2002) (of 287 case studies of markets for forest ecosystem services and the impact on the poor) concluded that the costs (social and other) of such schemes had rarely been assessed, 'the lack of attention to equity impacts of emerging payment schemes raises a number of concerns' (p. 5). In another study, Syme et al. (1999a) concluded that people in Australia have consistently rejected using water markets over ten years and 'effectiveness will depend on the community's agreement on the rules that underpin the market. This acceptance is likely to depend on fairness judgements' (p. 68). They also applied principles to other cultures (for example, Germany, see Syme et al. 1999b). In another study it was concluded, 'Whatever the case may be, equity considerations directly affect the acceptance of MBIs by the constituency' (Schilizzi 2003, p. 29). Therefore, apart from ethical reasons, the adherence to an equitable framework for MBIs may determine whether or not stakeholders will participate in these markets.

Equity does not only relate to what is perceived as being 'fair' between individuals although this may be a critical component of whether or not individuals will take part in such a scheme. There is also the need for consideration of fairness between humans and non-humans, for example, the allocation of water between human needs and environmental needs. A related issue is the practice involved in such market-based schemes of restricting access to public goods by privatizing these goods in order to allow trading to take place. Questions arise as to how far we should go in such schemes and who decides if this is to happen? Also, if all public goods were privatized, what would be the implications? These issues are all linked to equity and fairness considerations. The problem involved in assessing such issues is one of defining what is equitable and fair and how do we go about measuring equity and fairness.

In this chapter we review the literature on equity issues involved in PES and MBI schemes and establish a framework for the consideration of equity and fairness in such schemes, designed to protect and enhance ecosystem services. We do this by establishing a conceptual framework for measuring equity, fairness, and justice issues and by setting up a methodology for assessing such issues involved in the use of these instruments. The aim of the chapter is not to provide a definitive study on equity issues in PES instruments but to raise issues concerning equity that should be incorporated into these schemes. We start by reviewing some of the existing schemes, particularly those that have tried to address income equity (pro-poor schemes) and raise some important issues that address efficiency versus equity concerns. We then provide a conceptual framework and a methodology for adequately incorporating equity considerations into the design of such schemes on a case-by-case basis.

Is Equity Important?

Issues of equity and fairness have been widely debated in the PES literature. Schilizzi (2003) clearly states that while PES schemes are not designed for equity reasons, if equity is not taken into account and a scheme or policy is deemed to be inequitable, it will simply not be implemented.

Alix-Garcia et al. (2004) provides a broad overview of the literature on PES schemes. Most theoretical papers focus on targeting (Ribaudo 1989; Wu and Babcock, 1996; and Wu et al. 2000); a small group of empirical papers analyses farmers' willingness to participate in conservation programmes (Parks and Schorr 1997; Dupaz 2003), and there is a variety of case studies (Pagiola et al. 2002; Aylward and Togenetti 2002; Hernandez et al. 2003). Equity considerations are discussed in the case studies and in theoretical considerations of effectiveness and efficiency.

However, authors who do focus on equity as an issue admit that it is difficult to incorporate into an economic framework. Raymond (2003) claims that while equity-based norms have a significant influence over environmental decision making (citing evidence that a number of US PES schemes nearly failed because of equity-based disagreements), there is still a reluctance to consider equity explicitly in studies of public policy. Schilizzi (2003) explains that this reluctance stems from a multiplicity of competing equity principles.

Tables 2.1 and 2.2 illustrate how different notions of equity affect PES schemes. Table 2.1 shows that different notions of equity affect how it is viewed in PES schemes. Table 2.2, focusing on carbon trading schemes specifically, divides equity into three elements of access, legitimacy, and outcome. These tables are representative of the equity elements discussed in the literature and, taken together; provide an outline for considering equity in PES schemes.

The second section adopts the three elements of Table 2.2: access, outcome, and legitimacy which correspond to the questions of equality of opportunity, equality of outcome, and process in Table 2.1 to discuss whether PES schemes benefit the poor. The issue of positive discrimination outlined in Table 2.1 is a focus of the fourth section.

Table 2.1:   Different definitions of equity imply different questions

Equality of opportunity	What determines access to a PES scheme?
Equality of outcome	Are the poor equally likely to benefit as the rich?
Positive discrimination	Are schemes designed to benefit the poor rather than the rich?
Process	Do the poor participate in scheme design?

Source: Grieg-Gran (2004), online.

Table 2.2:   Three Elements of Equity in the Context of the New
Carbon Economy

Equity in access	Equity and legitimacy in institutions and decision making at all scales	Equity in outcome
Depends on information, communication, and knowledge; and the way institutions operate at different scales. Ease of access will determine participation and benefits from project outcomes	Concerns the way in which projects and rules operate and whether all stakeholders are able to have a voice in the project. Equity will not only be about participation but about inclusion and negotiation of competing views.	Concerns the way project outcomes impact the different stakeholders. The impacts will be conditioned and partially determined by access and decision making, but are primarily about who gains and who loses in terms of the distribution of project costs and benefits

Source: Brown and Corbera (2003).

How PES Impacts on Equity

It is as yet unclear how PES impacts equity (Schilizzi 2003; Grieg-Gran 2004). The diversity of the case studies shows mixed evidence because impacts on equity are design and context-specific (Grieg-Gran 2004). However, most authors agree that equity is an important consideration when it comes to PES schemes. Langeweg (1998) points out that PES schemes address equity and sustainable development issues because they can generate incomes for developing countries (especially in terms of climate change mitigation). Wilson and Howarth (2002) believe that equity considerations must be written into PES programmes because the allocation of ecosystem services directly affects many people and raises normative questions about social equity.

One equity implication of PES schemes is whether they will change existing power structures (either positively or negatively) in the control of essential ecosystem services. There is some disagreement in the literature. Jenkins et al. (2004) claim that creating markets for ecosystem services may fundamentally change the distribution of rights and responsibilities for ecosystem services. However, the International Institute for Environment and Development (IIED 2002) claims that markets generally reflect existing power structures. This view is supported by Corbera and Adger (2004) who say that most markets for ecosystem services are more likely to reinforce existing power structures and inequalities in access to resources. The UN Environment Programme (UNEP 2005) explains that in poor countries, the transfer and use of ecosystem services are usually done through non-market channels. Thus, the general consensus is that while existing inequalities in access and rights over essential ecosystem services may theoretically be affected by PES schemes; in most cases, they are not. This, in itself, is an equity issue and perhaps PES schemes should be designed to change existing power structures.

EQUITY CONSIDERATIONS IN PES SCHEMES

Do PES Schemes benefit the poor?

Most equity debates in the literature are framed around whether PES schemes benefit the poor. One of the reasons for government enthusiasm over PES schemes is their potential to benefit the poorest members of society. In fact, in many case studies, it is simply assumed that PES benefits the poor. However, the reality is somewhat different. A review of case studies (mainly from Latin America) shows that PES schemes can benefit the poor only if they are designed specifically with that goal in mind and Wunder (2005) warns that the small scale of most PES schemes constrains poverty alleviation.

Access

Tables 2.1 and 2.2 mention access as one of the elements of equity, and access is the first and most important hurdle for the poor to overcome. IIED (2002) outlines key constraints that limit the poor from accessing PES schemes. They are as follows:
1. lack of property rights over land and related environmental services;
2. inadequate technical and market-related skills;
3. poor market information;
4. lack of market contacts;
5. inadequate communication infrastructure;
6. inflexible contract design; and
7. lack of access to start-up capital.

The top constraint—lack of property rights over land is a key issue (Grieg-Gran 2004; Wunder 2005; Rosa et al. 2004; WWF 2006). The poorest of the poor are mostly landless peasants who either occupy land illegally or do not have formalized titles to the land they farm. Lack of land rights automatically preclude participation in most PES schemes. However, there are local PES schemes in Costa Rica which have a much more flexible criteria, allowing all those who work and live on the land (that is, not just landowners) to qualify for payments (Rosa et al. 2004).

The Costa Rican government set up a national PES scheme in 1995, which was designed to encourage forest protection and management by paying forest owners for four environmental services: carbon, biodiversity, watershed management, and landscape beauty (see Box 2.1 for its results). The scheme initially did not have any pro-poor mechanisms built into it but it was assumed that it would have a positive impact on poverty (Miranda et al. 2003). Because it was viewed partly as a conservation and partly as a social welfare programme, poor households that depend on other government benefit schemes and small landholders who were given land under the Agrarian Development Institute Programme were not eligible for PES payments, even if the land they owned satisfied the criteria (Miranda et al. 2003). The scheme was designed to conserve land at risk of deforestation, therefore, initially landholders who practiced agroforestry were excluded from payments. However, this exclusion was lifted in 2003 due to pressure from small landholders and indigenous groups (Rosa et al. 2004). This idea of leaving land idle also discouraged many smallholders from joining as they used their forests to grow shade coffee or as shelter for cattle. Setting land aside for conservation reasons was not feasible for them and prevented them from joining the PES scheme (Miranda et al. 2003).

The same problem is found in the developed world. The United States runs an Environmental Quality Incentives Programme which pays subsidies to encourage specific activities, such as nutrient management, fertilizer management, integrated pest management, irrigation management, and wildlife management. A review of the scheme found that 61 per cent of the US$ 22 billion paid out was received by 10 per cent of the farms, indicating that big landowners get a disproportionate share of the payments at the expense of smaller (and most likely poorer) landholders (Kumar 2005), reflecting the situation of their counterparts in the developing world.

Outcome

When the access barrier is removed, the poor may still not reap as many benefits from participation in a PES scheme as the relatively well-off. First and foremost, poorer participants face higher transaction costs (Wunder 2005; Grieg-Gran 2004; Miranda et al. 2003; WWF 2006). This was the case in Costa Rica where the requirements of leaving land idle while the application was in progress and substantial travel to obtain necessary documents made the scheme unattractive to many poor landholders (Grieg-Gran 2004).

Studies from Costa Rica indicate that the PES scheme brought financial rewards. Overall, the scheme increased household disposable income by 15 per cent, resulted in higher levels of investments on the farm, and contributed to some job creation (through the hiring of occasional workers) (Miranda et al. 2003). A small survey of PES participants living below the poverty line in the Oca Peninsula found that the scheme lifted half of them above the poverty line and became the primary household cash income source in 44 per cent of cases (Wunder 2005). However, in another area—the Virilla watershed, studies found that the participating landowners were mainly wealthy, well-educated, and did not directly live off the land (Miranda et al. 2003).

Box 2.1: The PES Experience in Costa Rica

The 1995 PES Scheme in Costa Rica emphasizes global environmental services provided by forests (particularly biodiversity conservation and carbon sequestration) and is primarily funded from a domestic tax on fossil fuels.

Between 1997 and 2002, the programme covered more than 300,000 hectares and total payments exceeded US$ 80 million with 70 per cent going for forest protection. The scheme contributed to the conservation of 16,500 ha of primary forest, the sustainable management of 2000 ha, and reforestation of 1,300,000 ha. By promoting live fences as well as sustainable agriculture and livestock practices, the scheme decreases the chances of land use conversion.

The scheme also had non-tangible effects by strengthening the process of institutional innovation, de-bureaucratization, decentralization, promoting voluntary agreements to improve the environment, and organizational and community innovation as well as fostering inter-institutional cooperation.

The scheme initially suffered from a number of equity issues that discriminated against the participation of the poor, but it evolved over time to address most of them.

Source: Miranda et al. 2003; Rosa et al. 2004.

Legitimacy

A third element of equity is the equity and legitimacy of institutions and decision making processes. The question here is whether the poor have a voice in designing PES schemes. This hinges on the issue of political power held by the poor. The World Wide Fund (WWF) for Nature points out that the poor lack skills, knowledge, and resources for participating in emerging markets. They also have little voice in the development of the markets and thus risk being marginalized from market benefits (WWF 2006). Janssen and Padilla (1999) find this is the case in the Philippines where mangrove areas are destroyed due to expansion of aquaculture (the conversion of swamps into fish ponds). The local population would support a PES scheme as the mangrove areas provide them with important environmental services (such as firewood, fish, and protection from floods). But the owners of fish ponds are wealthy individuals who do not live in the area nor employ local people to care for them and, therefore, are not directly benefiting from environmental services provided by the mangroves. Thus, they are not interested in their preservation.

Wunder (2005) says that PES schemes have a devastating impact on the landless poor who are engaged in environmentally degrading activities. This was a finding in Costa Rica where landowners considered security against squatters to be a major benefit of the PES programme (Miranda et al. 2003). Squatters are the poorest of the poor and, although squatting is illegal, they were hurt by the scheme. The landless poor are also likely to be employed in environmentally destructive activities such as logging, firewood and charcoal making, extracting non-timber forest products or they may be farm hands hired for clearing land and for cultivating converted soils (Wunder 2005). Any conservation scheme will thus hurt their interests by limiting their already meagre income. Another problem that is brought to light by Rosa et al. (2004) is that PES schemes can induce powerful outside interests to establish 'new' private property rights over resources previously managed by poor communities.

Corbera and Adger (2004) studied a carbon sequestration PES scheme in Chiapas, Mexico. They found that the project ignored internal conflicts and, therefore, reinforced existing unequal power relations within the community. In particular, landless families and women in general became excluded from project benefits because the carbon project ignored their role as carers of home gardens and focused on male-dominated tree planting.

The Distribution of Environmental Assets

The distribution of ecosystem services can have serious implications for equity. Some types of land are easier to conserve than others and this would make them better value for money in terms of PES, indicating that owners with land which, due to some environmental particularity, presents difficulties for conservation, would miss out in a PES scheme, regardless of their intentions to conserve. Two examples from Australia illustrate this.

The Australian state of Victoria has put to trial a successful auction system called BushTender. In BushTender, landowners identify actions and management strategies designed to preserve native vegetation on their lands. Then they prepare a bid, with the assistance of a government field officer which specifies management actions to be taken and their cost. The bids are then submitted to the Victorian Department of Natural Resources and Environment. The department chooses to fund those bids that present the best value for money (Stoneham et al. 2002). In essence, the system, while logical, is unfair because the price depends on the cost of provision of the service. In other words, two people providing the same service will be paid differently if their cost of producing the same service is different (Eigenraam et al. 2002). Therefore, landowners whose lands might make conservation difficult would struggle to compete against those whose land does not pose problems. However, the auctions are voluntary, so owners who would be disadvantaged are not forced to participate (Eigenraam et al. 2002). Despite this theoretical unfairness, anecdotal evidence suggests that landowners see the system as fair because the assessment system is relatively objective (Eigenraam et al. 2002). Hailu and Schilizzi (2003) conducted a hypothetical model of repeated auctions and found that repeated auctions (as opposed to the one-off system currently in use) would make the system inefficient and unfair due to information rents being extracted by winning bidders.

Flügge and Schilizzi (2003) examine a greenhouse gas restriction policy and how it could affect certain agricultural regions in Western Australia. They compare two different agricultural regions and how these would fare if a national tax on the amount of CO_2 equivalents emitted were applied. The Great Southern Region is livestock-dominant and, due to agro-climatic differences has very few options of switching to less CO_2-intensive production. The second region is the Eastern Wheatbelt Region which is crop dominant. Results of simulations showed that if the tax was in the order of $ 50 per tonne of emissions, farms in the Great Southern Region would go bankrupt while those in the crop-dominant region would survive (Flügge and Schilizzi 2003). A lower tax would still adversely affect one region more than the other. Therefore, the price for reducing harmful emissions nationally will be paid by one region more than another due to factors beyond the control of the people who would suffer.

EFFICIENCY VERSUS EQUITY

While equity is an important consideration in PES schemes, the ultimate goal is conservation, not fairness or poverty alleviation. As Schilizzi (2003) states, their raison d'être is economic efficiency, not distributional equity. Wunder (2005) takes up this argument by saying that from an efficiency point of view; only those who constitute a credible threat to the provision of environmental services should be paid. We have already seen above how this creates an unfair situation whereby those who combine conservation with income generating activities (such as agro-forestry or shade-grown coffee), are excluded from PES schemes. The efficiency argument states that these individuals (who are doing the right thing by taking care of the environment) are already receiving an income from the environment and limited PES funds should go to those who are doing the wrong thing by destroying the environment.

In order to maximize efficiency, PES schemes have been concentrating on single environmental services (such as carbon sequestration), sometimes at the expense of other ecosystem services; and giving priority to simplified, large-scale ecosystems, preferably controlled by a few people (that is, a few big landowners as opposed to many small landowners), so as to reduce transaction costs (Rosa et al. 2004). This has had adverse or devastating effects on poor and marginalized rural communities. Alix-Garcia (2004 et. al.) compared three types of PES schemes and found that the most egalitarian is also the least efficient.

The reason why most PES schemes insist on formal land title is also based on efficiency. PES payments are made for limiting resource use. Those without formal land rights cannot stop external agents from occupying the land and harvesting its resources (Wunder 2005).

PES schemes reward people financially for limiting resource use. However, in many instances resource use is already illegal (for instance, hunting wild animals, harvesting firewood, and deforestation) (Wunder 2005). If resource users who threaten the environment because of illegal activities receive payments to induce them to stop; is this not a de facto endorsement of crime? As far fetched as this might sound, deforestation is illegal in Costa Rica and the national PES scheme described here is in fact paying landowners not to cut down their trees, something they should not be doing anyway.

A similar national PES scheme in Mexico—the Payment of Hydrological Environmental Services (PSAH scheme—see Box 2.2) is designed by the federal government to pay participating forest owners for the benefits of watershed protection and aquifer recharge in those areas where commercial forestry is not currently competitive. However, most of the deforestation in Mexico occurs illegally, and therefore, the government scheme is again paying to stop illegal activities. Furthermore, forest owners with sustainable timber operations were excluded on the basis that they already benefited from the environment and were unlikely to destroy it. This decision was strongly challenged by timber and coffee producers who argued that it was unfair. They had for years conserved their forests by using them in a sustainable way and therefore they deserved payments more than those who failed to take responsibility for the environment (Muñoz-Piña et al. 2005).

In the end, a large share of participating forests were those which had some form of sustainable forestry activities—an outcome that was definitely fair and equitable (as the programme was found to help the poor and marginalized—see Box 2.2) but it was inefficient as payments were directed at forests that were unlikely to have been cut down in the first place (Muñoz-Piña et al. 2005).

Alix-Garcia (2004) makes a counter-argument in favour of paying to stop illegal activities. She claims that when the law is enforced, deforestation does decrease but this also eliminates income generation opportunities in forests of low commercial value. Therefore, when forest owners are poor, enforcing the law protects the environment at a cost of increasing poverty. There is thus a clear trade-off between equity and efficiency and poverty reduction and environmental protection.

It is important to note that not all PES schemes limit resource use. They can in fact be asset-building, for example, when trees are planted in degraded landscapes. Such activities can trigger a net expansion in rural jobs and benefit unskilled rural labour, thus alleviating poverty (Wunder 2005).

Box 2.2: The Payment for Hydrological Environmental
Services in Mexico

Mexico is generally considered to have the second highest
deforestation rate in the world. It suffers from soil erosion and
increasing water scarcity, problems associated with forest loss. It is
among the most biologically diverse countries in the world, with
first place in reptilian diversity, third in bird diversity, and fourth
in mammal diversity. Eighty per cent of the country's forests are
located in ejidos (community managed areas).

The PSAH scheme consists of direct payments to landowners
with primary forest cover (forests in good state of conservation)
given at the end of the year, once it has been proven that they were
not deforested.

In 2003, more than 900 applications were received offering close
to 600,000 hectares. Only 271 forest owners were selected
incorporating 127,000 hectares into the programme. In 2004,
thanks to Congress support, the budget was increased by 50 per
cent. The number of applicants grew to 960, of which 352 new
participants were chosen with approximately 180,000 hectares.

A first and positive result is that, despite not being an explicit
criterion, 72 per cent and 3 per cent of PSAH payments in 2003
and 2004 respectively went to forests where population centres have
high or very high marginalization. Between 2003 and 2005, satellite
images showed that less than 0.1 per cent of the nearly 300,000
hectares paid by the programme was deforested. And those areas
that were lost suffered from unintentional and very difficult to
control forest fires, not land use changes.

Source: Alix-Garcia (2004); Muñoz-Piña et al. (2005).

The equity versus efficiency dilemma is also apparent in the
developed world and illustrated by a well-established scheme in
America—the Wetland Mitigation Bank. This scheme demands that
developers who want to develop (that is, turn into a mall or housing
estate) land containing wetlands must first either protect an existing
wetland or create a new one somewhere else before destroying one.
The aim is not to have a net loss of wetlands but the rules are flexible
as to wetland type and quality. This process has resulted in the
destruction of wetlands in urban/suburban areas and establishment
of wetland mitigation banks (large tracks of contiguous wetlands)
in rural/sparsely populated areas. The outcome is that wetlands have

moved out of areas where they may provide services to urban populations. The process has involved a trade-off between concerns over equity in terms of who has access to wetland services and economic efficiency (it is more efficient to create wetlands in sparsely populated areas where the people cannot enjoy their services) (Salzman and Ruhl in Pagiola et al. 2005).

MAKING PES MORE EQUITABLE

So far the literature has shown that: (i) equity plays an important role in PES schemes; (ii) PES schemes are meant to maximize efficiency and not equity; (iii) the case studies show that schemes have been adjusted to make them more equitable and less efficient. This indicates that in order to be applicable in the real world, PES schemes should become more equitable, even if this decreases their efficiency.

Much has been written on how to make PES more equitable. For example, both the IIED (2002) and WWF (2006) researched how to make PES more pro-poor. Their findings are detailed in Table 2.3.

Table 2.3:   Mechanisms to create pro-poor markets

WWF		IIED	
Mechanism	Description	Mechanism	Description
Formalize forest service property rights held by poor people	Formalization of natural resource rights will give marginalized groups control over, and rights to, returns from environmental service sales	*Property Rights*	Property rights over land and related environmental assets must be assigned in ways that respect customary arrangements and are equitable
Define appropriate commodities	Simple and flexible commodities that can be self-enforced, that fit with existing legislation and that suit local livelihood strategies need to be developed in poorer areas	*Market Participation*	Strengthening capacity for market participation, for example, through training and education
Devise cost-effective payment mechanisms	In areas where regulatory capacity is weak, trading skills in short-supply and market infrastructure underdeveloped, simpler	*Market Support*	Support through the provision of market information, advice, a contact point for buyers and sellers, and

(contd...)

Table 2.3 (contd...)

WWF		IIED	
Mechanism	Description	Mechanism	Description
	payment mechanisms are likely to be most effective		facilitation in the bundling of service contracts will reduce transaction costs
Strengthen cooperative institutions	Cooperation is critical in allowing poor landowners and service beneficiaries to share the costs associated with market participation. It is also essential for achieving a minimum level of supply or demand, thereby permitting market participation	*Start-up Capital*	Improving access to start-up capital so that poor individuals can make necessary investments in market participation
Invest in training and education	Training in marketing, negotiation, management, financial accounting, contract formulation, and conflict resolution are important prerequisites for effective participation. Technical skills relating to forest management for environmental services are also needed.		
Establish a market support centre	To improve poor people's ability to participate in emerging markets, a central market support centre could offer free access to market information, a contact point for potential buyers, sellers, and intermediaries, and an advice bureau to support the design and implementation of contracts		
Improve access to finance	Where finance is needed to negotiate and conclude environmental service deals, the government may have a role to play in supporting access to funds		

Source: WWF (2006); IIED (2002).

Rosa et al. (2004) provide a comprehensive review of Latin American PES schemes and provide a useful summary of lessons learned from each country. These lessons correspond well to the findings of the WWF and IIED.

Both lists emphasize non-financial support to the poor. Information, training, and education are important along with the improvement of institutional capacity. This type of support deemed essential by two of the foremost international environmental organizations lies in the arena of international development and national education policies. In other words, environmental agendas must create partnerships with social development and poverty alleviation. There is no escaping the fact that non-environmental considerations such as equity are essential to solve environmental problems.

In fact, WWF has created an equitable PES scheme which pursues a balanced approach towards poverty reduction and sustainable management of environmental services. The WWF scheme differs from traditional PES approaches in its focus on achieving equity and an equitable process of implementing the scheme (WWF 2006). The WWF's scheme is currently under trial. It has partnered up with CARE International to implement the Equitable Payments for Watershed Services (PWS) in 10 selected watersheds in Asia, Africa, and Latin America (WWF 2006). This project is in its early stages and so far no evaluations are available.

THE WAY FORWARD

Worldwide, market-based measures to control pollution or environmental degradation are becoming more and more popular as compared to the traditional command and control measures. Andersen and Sprenger (2000, p. 9) looked at the rise of the use of market instruments in Organisation for Economic Cooperation and Development (OECD) countries over several decades and concluded that 'comparing the data for the eight best-documented countries, the number of economic instruments in use in 1992 was 25 per cent higher than in 1987. If the number brought into use in 1993 is also taken into account, the increase is nearly 50 per cent'. This brings the total number of economic instruments to around 225 for the OECD countries in 1993.

In essence, PES schemes provide market signals to encourage certain reactions from market participants and are often credited with being more efficient than other methods such as those based on 'command and control' or 'polluter pays' principles. Market schemes may use, for example, trading mechanisms, auctions, and price signals (in the form of subsidies and taxes) to change behaviour (Murtough et al. 2002). Different instrumentism however have different implications for equity and fairness considerations (Schilizzi 2003) and must, therefore, be considered on a case-by-case basis.

The rate of taking up of PES programmes and the types of instruments used has differed in different countries. Some researchers claim that their implementation rate has been linked to different views on equity in different countries. For example, the acceptance of some market instruments (particularly tradeable permits) as an environmental policy measure has been more popular in the US and Australia than in Europe. There may be other reasons, however, such as institutional structures that are already in place to allow the implementation, that may play a role in the different rates of use of market instruments in different countries:

It may be that Europeans have a different sense of equity than Americans have—they may be referring, because of historical reasons or otherwise, to different equity principles. The existence of institutional structures which allow the introduction of MBIs in the USA more easily than they do in Europe also raises questions as to the nature and pace of institutional change, and how perceptions of equity influence this change (Schilizzi 2003, p. 29).

In Australia, trading in water entitlements has been undertaken in specific regions since the 1980s in response to continued environmental degradation as a result of limited environmental flows and the fact that the major user of the water resource is the irrigation industry. Equity issues involved in trading water permits has long been discussed (see, for example, Syme and Fenton 1993; Syme and Nancarrow 1992) although little has been done at the policy and implementation levels to address these equity concerns. Reasons of 'lack of fairness' have often been cited as crucial issues in determining whether farmers will participate in such instrument schemes.

At present in Australia a major commonwealth programme (The National Market-Based Instruments Pilot Programme) has been initiated to explore the advantages and disadvantages of such instruments with both experimental and on-ground trials by providing $5 million worth of funding over 21 priority regions. Although not explicitly mentioned as an area of investigation for these pilot studies, issues of equity and fairness need to be investigated for the afore mentioned reasons of possibly deterring participation if such concerns are not addressed and also because as a policy instrument, the ability of such instruments to achieve sustainability criteria will be partly determined by addressing the equity concerns of the participants of such MBI programme.

CONCEPTUAL FRAMEWORK

When trying to build a conceptual framework upon which practical studies of assessing equity considerations can be undertaken, the question remains: what do we mean by equity and how do we measure it? Amartya Sen defines equity as 'equality of something' where 'something' includes tangible and non-tangible resources. The dictionary definition (ignoring the definition related to the business accounting term) states that equity is *the state, quality, or ideal of being just, impartial, and fair* or *something that is just, impartial, and fair*. A relevant legal definition also exists, which is *justice applied in circumstances covered by law yet influenced by principles of ethics and fairness*. The definitions, therefore, suggest that something that is just and fair is also considered equitable, which leaves us with the problem of a single and exact definition of equity that could be used for measurement as justice and fairness will have many different interpretations for many different people. Table 2.4 summarizes the many ways that justice has been assessed in various environmental studies and suggests that the interpretation and measurement of equity is a situation-specific phenomenon,[1] best dealt with and measured by those who would be affected by the situation being assessed—the stakeholders.

Generally, such issues can be divided into those of 'procedural justice' or a *process* that will ensure a fair and just outcome and those of 'distributive justice' which is concerned with the final *allocation* of rewards and responsibilities, regardless of the process.

---

[1] Syme et al. (1999a) refer to *universal* fairness criteria versus *situation-specific* criteria.

Table 2.4: Review of principles and criteria linked with procedural and distributive justice*

Subject	Principle	Description	Level of equity	Source
Distribution of allocative resources				
Income	–	Household income	Intragenerational	Brown (2003)
Income	'No envy' principle	It conveys the ideal of equal opportunity of consumption and defines a situation where no agent would prefer someone else's consumption bundle to his own (Diamantaras and Thomson 1990). Thus, its requirement is that every active agent should bear the same cost or enjoy the same gain (Varian 1974).	Intragenerational	Ikeme (2003)
Income	'Just deserts' concept	This option seeks remedies that are proportionate to the weight of the injustice. So remedies for injustice should not engender a secondary inequity.	Intragenerational	Ikeme (2003)
Income	Total equality concept	It argues that everyone should have the same income, i.e. the bottom 10 per cent of the population should receive 10 per cent of the income (Le Grand et al. 1976; Stymne and Jackson 2000).	Intragenerational	Ikeme (2003)
Income	Minimum standard or basic need approach	It is concerned only with the poor in the society and argues that nobody's income should fall below a certain minimum level (Le Grand et al. 1976; Stymne and Jackson 2000).	Intragenerational	Ikeme (2003)
Negative impacts	Relinqui-shment	Not to carry out the project: If a project causes irreversible harmful effects to future	Intergenerational	(Padilla 2002)

(contd...)

Table 2.4 (contd...)

Subject	Principle	Description	Level of equity	Source
		generations and these cannot be avoided or compensated, it should be considered outside the choice of possibilities.		
Negative impacts	Precautio-nary and control measures	This option also implies the application of the inalienabi-lity rule (the inalienability rule involves a much more restrictive use of the power in present decision making. This rule implies that the present cannot modify certain rights of future generations). If the modification of the structure of rights that the original project would imply is avoidable (for example, enhancing security systems) and it is still profitable, this option is more appropriate than the first one.	Intergenerational	Padilla (2002)
Negative impacts	Compen-sation through an associated project	In some projects it is possible to compensate the harmful effects on future generations through an associated project (for example, reforesting) (see Markandya and Pearce (1988).	Intergenerational	Padilla (2002)
Negative impacts	Financial compen-sation	This option would clearly modify the composition of the capacity bequeathed to future generations. There should not be doubts about the possibility of substituting the diminished resources and of establishing an investment fund allowing this future compensation. See Costanza and Perrings (1990).	Intergenerational	Padilla (2002)

(contd...)

Table 2.4 (contd...)

Subject	Principle	Description	Level of equity	Source
Negative impacts	Compens-ation	Identifying compensation measures for those adversely affected by implementation of a project.	Intragenerational	Orlando (2002)
Negative impacts	Minimi-zation	Devising strategies to minimize negative impacts on people's lives.	Intragenerational	Orlando (2002)
Welfare	Soverei-gnty	Equalize net welfare change across nations	International	Rose (1998)
Welfare	Vertical	Welfare gains should vary inversely with national economic well-being	International	Rose (1998)
Welfare	Compen-sation	Distribute permits so no nation suffers a net loss of welfare	International	Rose (1998)
Welfare	Access for poorest	Forest resources access to poorest households	Intragenerational	Brown (2003)
Welfare	Commu-nity invol-vement	Number of local people parti-cipating in project activities and who perceive benefits	Intragenerational	Brown (2003)
Welfare	Capacity building	Investment in education, health services, and capacity building	Intragenerational	Brown (2003)
Distribution of authoritative resources				
Property rights	Sovereignty	All nations have an equal right to pollute and to be protected from pollution → Distribute permits in proportion to emissions	International	Rose (1998)
Property rights	Egalitarian	All people have an equal right to pollute or to be protected from pollution → Distribute permits in proportion to population	International	Rose (1998)
Property rights	Ability to Pay	Mitigation costs should vary directly with national economic well-being → Distribute permits to equalize abatement costs	International	Rose (1998)

(contd...)

Table 2.4 (contd...)

Subject	Principle	Description	Level of equity	Source
Property rights	Initial allocation	Clarification of property rights	–	Brown (2003)
Procedural justice and legitimation				
Income	Merito-cracy	Inequality is accepted if every-one has had equal opportunity at initial allocation and differential is only accounted for by difference in effort and hard work	Intragenerational	(Konow, 2001) in Ikeme (2003)
Property rights	Consensus	The international negotiation process is fair → Distribute permits in a manner that satisfies the (power weighted) majority of nations	International	Rose (1998)
Property rights	Market Justice	Market is fair → Distribute permits to highest bidder	International	Rose (1998)
Welfare	Rawls' Maximin	The welfare of the worst-off nations should be maximized → Distribute largest proportion of net welfare gain to poorest nations	International	Rose (1998)
Procedural justice and signification				
Participation	Commu-nicating	Determining and communi-cating project boundaries	Intragenerational	Orlando (2002)
Participation	Commu-nity invol-vement	Involving people as much as possible in the project process	Intragenerational	Orlando (2002)
Participation	Commu-nity invol-vement	Involvement of community-based formal and non-formal organizations in project	Intragenerational	Brown (2003)

Notes: *Giddens structurization theory is used as the basic framework;
Signification (for example, shared understanding of ecosystem services);
Legitimation (norms/rules like definition of property rights over ecosystem services).

METHODOLOGY

A methodology has been developed here to allow the inclusion of stakeholders in the development and design of context-specific PES schemes to account for the issues raised in this chapter in interpreting and measuring equity, fairness, and justice. In this methodology, a suitable scheme can be developed with the aid of stakeholder input and utilizing an experimental economics framework along with stakeholder surveys. The procedure includes the following stages.

1. Survey of landholders—to assess demographics and attitudes.
2. Experimental economics—to test reactions to different schemes and aid design and learning.
3. In-depth survey—involving equity considerations.
4. Multi-criteria evaluation—to determine a favoured instrument.
5. On ground trial—to test and monitor the instrument.

An example of this methodology is now provided where the relevant stakeholders are farmers but the technique could be extended to other types of stakeholders as well.

In the first stage, the first survey is sent out to all farmers in the region under study and will generally be assessing the types of farms and farmers that reside in the region. The questions to be asked include the following.

- Farmers' thoughts on farming, including attitudes towards environmental conservation and their responsibilities, whether they are worried by the views of their neighbours and their attitudes towards new methods and techniques of farming.
- Farm type and the extent of the salinity problem on their farm.
- Farming practices as well as their use of technology such as computers.
- Cropping enterprises.
- Grazing enterprises.
- Demographics.

These questions can be used later to assess what types of farmers are willing to take part in a market instrument scheme. In the second stage, the experimental economics framework aids in testing the way farmers will react to different types of schemes given their existing knowledge and farm characteristics but the testing is done using a virtual trial of the instrument with the aid of computer facilities under 'laboratory' conditions. The experiments will aid in understanding of the schemes by the farmers that could be eventually used in on-ground trials. At the end of the experimental phase, farmers will have greater knowledge of the way such schemes work and the implications of each scheme for their own farming situation as well as their likelihood in taking part in a 'real' scheme. It will also aid them in understanding the specific issues of fairness and equity involved in such schemes and whether or not equity or fairness issues play any role in how the participants respond to different process designs. This should enlighten further work on what types of issues need to be investigated and addressed when designing certain types of PES schemes.

A second questionnaire is then given to the experiment participants to fill out after the experiments are conducted. This questionnaire is based on one that was developed to examine equity and fairness issues involved in water trading by farmers in Australia (Syme and Nancarrow 1992). Individuals are asked to rank on a scale of 1 to 5, the degree to which they agree with various statements related to equity issues involved in the particular PES scheme. They are then asked to what extent they believe various schemes can be rated on a fairness scale and be prompted for further information if they believe certain schemes are unfair.

The next stage of the process is to engage farmers involved in the experiments, resource managers, agricultural extension officers, researchers, and other local stakeholders in developing three or four different schemes that will be evaluated using multi-criteria evaluation. The criteria to be used will include, for example: the likelihood of rent seeking processes in the scheme, equity criteria (developed using the results of the two surveys), efficiency criteria, transaction costs, the likelihood of moral hazard problems, and the likely participation rates for a particular scheme.

The final part of the framework uses the results to implement a scheme as an on-ground trial for a group of farmers in the area. Monitoring and self auditing are then included as important components of the on-ground trial.

Discussion

In this chapter we have tried to investigate several issues involved in the assessment of fairness and equity related to MBIs in general and some that are particular to certain types of schemes. For example, these may be related to the perceptions of fairness and equity by the participants, issues of access by participants to the market-based scheme as well as the procedures used to estimate and allocate such PES. Specific to this study is the issue of involvement of stakeholders in developing the scheme and to review and assess issues related to equity and fairness in the scheme in question. Such issues may, for example, be related to the degree of fairness involved when environmental problems are deemed to be the responsibility of private landholders; the design of this specific scheme (and who participates in this design) and how it should be perceived as being 'fair' by the community; how the process should deal with random events of weather changes (as floods and droughts will have serious impact on outcomes) in a 'fair' way; and how payments are deemed fair given that individual farmers start with different levels of environmental problems on their land that may result from past management, hydrogeology, and typography.

REFERENCES

Alix-Garcia, J., A. de Janvry, and E. Sadoulet (2004), 'Payments for Environmental Services: To Whom, for What, and How Much?', University of California at Berkeley', website: *http://www.are.berkeley.edu/ ~alix/finalPESsimulations.pdf*

Anderson, M. and R-U. Sprenger (2000), *Market-based Instruments for Environmental Management: Politics and Institutions*, Edward Elgar, UK.

Alyward, B. and S. Tognetti (2002), 'Valuation of Hydrological Externalities of Land Use Change: Lake Arenal Case Study, Costa Rica,' Land–Water Linkages in Rural Watersheds, Case Study Series, FAO, Rome.

Brand, D. (2002), 'Investing in Environmental Services of Australian Forests', in S. Pagiola, J. Bishop and N. Landell-Mills (eds) (2002), *Selling Forest Environmental Services: Market-based Mechanisms for Conservation and Development*, Earthscan Publications, UK.

Brown, E, and E. Corbera (2003), 'Exploring Equity and Sustainable Development in the New Carbon Economy', Climate Policy, Vol. 3S1 (2003), pp. S41–S56.

Corbera, E. and W.N. Adger (2004), 'The Equity and Legitimacy of Markets for Ecosystem Services: Carbon Forestry Activities in Chiapas, Mexico,' Paper presented at the International Association for the Study of Common Property Conference, Mexico, 2004, website: *http://dlc.dlib.indiana.edu/*

Costanza, R., C. Perrings (1990), 'A Flexible Assurance Bonding System for Improved Environmental Management', *Ecological Economics*, Vol. 25, pp. 55–7.

Dupaz, P. (2003), 'The Environmental Supply of Farm Households: A Flexuable Willingness to Accept Model,' *Environmental and Resource, Economics*, Vol. 25, No. 2, pp. 171–89.

Eigenraam, M., L. Strappazzon and G. Stoneham (2002), 'Sustainable Hardwood Production: Economic Principles for Allocating Public Resources', Report completed for the West Victorian Private Forestry Project, Victoria.

Flügge, F. and S. Schilizzi (2003), 'Greenhouse Gas Abatement Policies and the Value of Carbon Sinks: Do Grazing and Cropping Systems have Different Destinies?', Paper presented at the 47th Annual Conference of the Australian Agricultural and Resource Society, Fremantle, Western Australia, 12–14 February 2003.

Grieg-Gran, M. (2004), 'Pitfalls and Potential in Addressing Equity: How can MES Work for the Poor?', Global Systems Workshop: Emerging Payments for Ecosystem Services, World Conservation Congress, 18–19 November, 2004, Bangkok, website: *www.biodiversityeconomics.org/document.rm?id=701*

Hailu, A. and S. Schilizzi (2003), 'Learning in a "Basket of Crabs": An Agent-based Computational Model of Repeated Conservation Auctions', Paper presented to the international Workshop on the Economics of Heterogeneous Interactive Agents (WEHIA), Kiel, Germany, 31 May 2003.

Hernández, O., C. Cobes, A. Ortiz, and J.C. Méndez (2003), 'Valóracion Económica del Servicio Ambiental de Regulación Hindrica del Lado Sur de La Reserva de La Biosfera Surra des Las Munas, Gualemala,' Paper prepared for Foro Regional Sobre Sistemas de Pago por servicios Ambientales, Aréqupa, Peru.

IIED (International Institute for Environment and Development) (2002), *Equity for a Small Planet: Challenges and Ideas for the World Summit on Sustainable Development*. website: *www.iied.org/pubs/display.php?o=11024IIED&n=83&l=250&c=policy*

Janssen, R. and J.E. Padilla (1999), 'Preservation or Conversion? Valuation and Evaluation of a Mangrove Forest in the Philippines', *Environmental and Resource Economics*, Vol. 14, pp. 297–331,

Jenkins, M., S. Scherr, and M. Inbar (2004), 'Payment where it's due', *Environmental Finance,'* November, pp. 24.

Kumar, P. (2005), 'Market for Ecosystem Services', International Institute for Sustainable Development, website: *www.iisd.org/pdf/2005/economics_market_for_ecosystem_services.pdf*

Landell-Mills, N. and I. Porras (2002), *Silver Bullet or Fools' good? A Global Review of Markets for Forest Environmental Services and Their Impact on the Poor*, IIED, London.

Langeweg, F. (1998), 'The implementation of Agenda 21 "our common failure"?', *The Science of the Total Environment,* Vol. 218, pp. 227–38.

Markandya, A. and D.Pearce (1988), 'Sustainable Future, Natural environment and the social rate of discount,' *Project Appraisal,* Vol. 3, pp. 2–12.

Miranda, M., I.T. Porras, and M.L. Moreno (2003), 'The social impacts of payments for environmental services in Costa Rica', Environmental Economics Programme, IIED, website: *www.iied.org/pubs/display.php?o=9245IIED*

Muñoz-Piña, C., A. Guevara, J.M. Torres, and J. Braña (2005), 'Paying for the Hydrological Services of Mexico's Forests: Analysis, Negotiations and Results,' Instituto Nacional de Ecología, website: *www.ine.gob.mx/dgipea/download/draft_ecological_economics.pdf*

Murtough, G., B. Aretino, and A. Matysek (2002), *Creating Markets for Ecosystem Services,* Productivity Commission Staff Research Paper, Ausinfo, Canberra.

Pagiola, S., J. Bishop, and N. Landell-Nills (eds) (2002), *Selling Forest Environmental Services: Market-based Mechanisms for Conservations and Development,* Earthscan Publications, UK.

Parks, P.J. and J.P. Schorr (1997), 'Sustaining open space benefits in the northeast: An Evaluation of the Conservation Reserve Program', *Journal of Environmental Economics and Management,* Vol. 32, pp. 85–94.

Ribaudo, M. (1989), 'Targeting the Conservation Reserve Program to Maximize Water Quality Benefits, *Land Economics,* Vol. 65, No. 4, pp. 320–32.

Raymond, L. (2003), *Private rights in public resources: Equity and property allocation in market-based environmental policy,* Resources for the Future, Washington, DC.

Rosa, H., D. Barry, S. Kandel, and L.Dimas (2004), 'Compensation for Environmental Services and Rural Communities: Lessons from the Americas', Political Economy Research Institute, University of Massachusetts, Working Paper Series 96, website: *www.peri.umass.edu/Publication.236+M539bd54882c.0.html*

Schilizzi, S. (2003), 'Should Equity Concerns Impose Limits on the Use of Market-based Instruments?', in Proceedings of the Australian Agricultural and Resource Economics Society National Symposium on Market-based Policy Instruments, 2–3 September 2003, Canberra, Australia, RIRDC, ACT, Australia.

Slazman, J. and J.B.Ruhl (2002), 'Paying to Protect Watershed Services: Wetland Banking in the United States', in S. Pagiola, J. Bishop, and N. Landell-Mills (eds), *Selling Forest Environmental Services: Market-based Mechanisms for Conservation and Development,* Earthscan Publications, UK.

Stoneham, G., V. Chaudhri, A. Ha, and L. Strappazon (2002), Auctions for Conservation Contracts: An Empirical Examination of Victoria's BushTender Trial, Department of Natural Resources and Environment and Melbourne Business School, Victoria, Australia.
Syme, G. and D.M. Fenton (1993), 'Perceptions of Equity and Procedural Preferences for Water Allocation Decisions', *Society and Natural Resources*, Vol. 6, pp. 347–60.
Syme, G. and B. Nancarrow (1992), 'Perceptions of Fairness and Social Justice in the Allocation of Water Resources in Australia', A Report for the Land Water Research and Development Corporation, CSIRO Division of Water Resources.
Syme, G., B. Nancarrow, and J. McCreddin (1999a) ,'Defining the components of fairness in the allocation of water to environmental and human uses', *Journal of Environmental Management*, Vol. 57, pp. 51–70.
Syme, G., E. Kals, B. Nancarrow, and L. Montada (1999b), 'Ecological risks and community perceptions of fairness and justice: A cross-cultural model', *Risk Analysis*, Vol. 20, No. 6, pp. 905–16.
UNEP (2005), 'Concept Note: Creating Pro-poor Markets for Ecosystem Services', High-level Brainstorming Workshop, 10–12 October 2005, London, United Kingdom, website: *www.unep.org/dec/docs/ Concept_note_lon.doc*
Wilson, M.A. and R.B.Howarth (2002), 'Discourse-based Valuation of Ecosystem Services: Establishing Fair Outcomes Through Group Deliberation', *Ecological Economics*, Vol. 41, pp. 431–43.
WWF (World Wide Fund for Nature) (2006), 'Payments for Environmental Services: An Equitable Approach for Reducing Poverty and Conserving Nature', WWF, June 2006, The Netherlands, website: *assets.panda.org/ downloads/pes_report_2006.pdf*
Wu, J. (2000), 'Suppage Effects of the Conservation Reserve Programs', *American Journal of Agricultural Economics*, Vol. 82, pp. 979–92.
Wu, J. and B.A. Babcock (1996), 'Contract Design for the Purchase of Environmental Goods from Agriculture', *American Journal of Agricultural Economics*, Vol. 78, No. 4, pp. 935–45.
Wunder, S. (2005), 'Payments for environmental services: Some Nuts and Bolts', CIFOR Occasional Paper, No. 42, Jakarta.

3

An Institutional Analysis of Negotiation Support Systems for Watershed Environmental Services
A Case Study of the Bhoj Wetlands, Madhya Pradesh, India*

SUNANDAN TIWARI AND JAIME AMEZAGA

INTRODUCTION

There is a growing recognition of the importance of maintaining environmental services through protection and improved management practices. Environmental services include carbon sequestration, biodiversity conservation, landscape beauty, and watershed services. Typically, poor and marginalized rural communities reside in the areas from which these services are derived. A two-pronged approach of developing incentive-based mechanisms for protecting these services while simultaneously addressing the livelihood concerns of these poor communities is being increasingly experimented with in different parts of the world. This approach encourages communities whose actions could jeopardize the provision of environmental services, or the service providers, to appropriately alter specific practices in order to maintain the flow of these services. In return, communities that benefit from these services (service receivers) provide incentives, monetary or otherwise, to the service providers to continue with these improved practices. The processes through which this interaction between service providers and service receivers is developed and

*This chapter is based on research under the 'Low Base Flows and Livelihoods in India' project, funded by United Kingdom Department for International Development (DFID). The views are not necessarily those of DFID (R8171 Forestry Research Programme).

maintained is termed as negotiation support systems (NSS). This chapter specifically examines NSS for environmental services derived from watersheds.

Managing and maintaining watershed environmental services (WES) represents a collective action problem of significant complexity. As illustrated in the 'tragedy of the commons' (Hardin 1968), individual behaviour aimed at maximizing personal gains can jeopardize the sustainability of common resources, such as services provided by watersheds. To tackle this sort of 'selfish' behaviour, efficient institutions, which are a fundamental way for solving collective action problems, can reduce the uncertainty in the behaviour of individuals and create incentives towards greater levels of coordination and cooperation (Bravo 2002). However, achieving this level of coordination and cooperation is not a simple task. Bravo (2002) refers to Bates (1988), who points out that creating institutions to overcome a collective action problem is itself a collective action problem of a higher scale. Therefore, as critical as it may be to promote institutions for the management of WES, it is equally critical to have an improved understanding of the functioning of these institutions within their specific physical and social settings and of the NSS that facilitate transactions related to environmental services.

This chapter presents an analytical framework that demonstrates a potential methodology for analysing and designing successful NSS for the improved management of WES. The potential of this analytical tool is being tested through an ongoing action learning project aimed at developing incentive-based mechanisms for WES in the Bhoj Wetlands of Madhya Pradesh, India.

THE ANALYTICAL FRAMEWORK

The analytical framework draws on the Institutional Analysis and Development (IAD) framework (Ostrom, Gardener, and Walker 1994; Ostrom, 1990) and its adaptations (Fischer and Petersen 2004; Thomson 1992). Stated very broadly, 'the IAD framework seeks to explore and link the influences on three levels: constitutional, collective action, and operational' (Gibson 2005, p. 229). As per the IAD framework, three external factors, namely, the 'rules-in-use', the 'attributes of the community' and the 'attributes of the physical world' affect its 'conceptual core—the action arena—where actors interact in a social space, named action situation. The characters of actors and of the action situation define the arena, while activities, interactions, and exchanges among individuals inside the arena produce the outcomes of the institutional arrangements' (Bravo, 2002; p. 2). Within the IAD, the action arenas are viewed as dependent variables while the rules-in-use are the most important independent variables in institutional analysis, because the rules determine the behaviour of the actors (Mulholland and Shakespeare 2005).

While the IAD framework offers an approach to analyse incentives that guide the users' behaviour with regard to natural resources, it does not capture the different roles of multiple actors, such as public authorities, governments, and non-governmental organizations (NGOs), their interactions and relationships (Fischer and Petersen 2004). In the case of watershed management, these 'external' actors play a key role in implementing incentive changes and, therefore, need to be taken into consideration. This framework is adapted from the model proposed by Fischer and Petersen (2004) who have, in turn, drawn heavily on the work of Thomson and Schoonmaker Freudenberger (1997), Thomson (1992), and Ostrom (1990).

This framework adds a more elaborate second part to the IAD framework and consists of the following:
1. The first part provides an instrument to analyse the attributes and motives of the actual resource users' behaviour (analysis of local factors).
2 The second part serves to describe the activities and incentive measures provided by external actors, that is, other than the local communities (analysis of external actors).

The framework analyses each of these two parts independently and also examines how the activities and inputs provided by external actors influence the attributes and motives of the communities and their NSS for WES (see Figure 3.1). The framework can be used ex post to evaluate existing examples and to derive respective lessons learned, as well as ex ante to design appropriate strategies and activities (Fischer and Petersen 2004). It focuses on three pivotal issues that would help assess the efficacy of these institutions in managing and maintaining the related environmental services, namely, (i) identification of the key incentives for improving watershed management; (ii) nature of the negotiations; and (iii) sustainability of the mechanism.

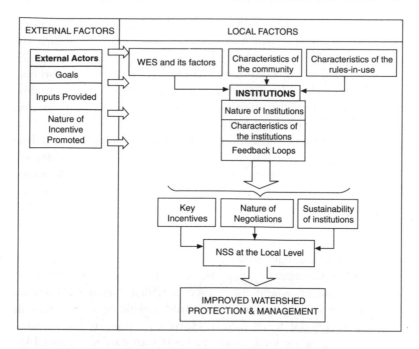

Fig. 3.1:   Framework for Institutional Analysis

As depicted in Figure 3.1, there are two parts to this framework that influence negotiation support systems at the local level: 'external factors' and 'local factors'. External factors create a 'push effect', while local factors relate to the 'pull effect'. Therefore, efforts are most productive when both these 'effects' function in a complementary manner. In the case of 'local factors', the action arena relates to a particular situation in which there exists an NSS or potential for developing such a system. Therefore, 'local' refers to the universe within which NSS for WES can or do exist. The external factors are typically determined by the goals of the external actors, their inputs into a local system, and the nature of incentives that they aim to promote.

Institutions are the building blocks for NSS and, therefore, play a pivotal role in developing such systems. In this case, 'institutions at the local level' represent the 'actors' and the social conditions within which these institutions function represents the 'action situation'. As per the IAD framework, the actors and the action situation define the 'action arena' which is influenced by three factors—the 'attributes of the physical world' which in this case corresponds to 'watershed environmental services and their characteristics', 'attributes (or characteristics) of the community', and the '(characteristics of) rules-in-use'—at the local level. In this framework, three broad attributes of institutions are to be investigated as shown in Figure 3.1 and are explained in the following section.

The composite of 'institutions' and its influencing factors, viz. WES, community characteristics, and rules-in-use, are further impacted by 'external factors' (explained above). These external factors can impact the nature, characteristics, and functioning of institutions, as well the watershed services through water harvesting interventions, for example, the characteristics of communities through social mobilization activities, and the rules-in-use by introducing new or modified resource management systems.

Therefore, the emerging 'negotiation support system' for improved watershed protection and management, which corresponds to the 'outcomes' in the IAD framework, would depend on the combined effect of local and external factors. However, a point to note is that while it is critical for local factors to be at least partially favourable, the magnitude of the impacts or role of 'external factors' can vary from case to case. For example, it is possible that stakeholders at the local level initiate an NSS process on their own with some basic support from external agencies. In this case, the role of 'external factors' in developing an NSS at the local level would be low, albeit, they may play a greater role at a later stage in maintaining or expanding the developed system. On the other hand, the role of external actors can be substantial as is typically the case with government-led watershed development programmes. The framework will, therefore, not only assist in determining the local and external factors that support negotiation systems for improved WES, but is also flexible to accommodate situational variations.

Applying the Framework

The parameters for analysis within each of the two parts of the framework are described here.

Analysis of Local Factors

The first part of the framework assists in identifying and analysing the local factors in question and consists of the following:

- Identification of the service (for example, water quality and quantity, silt control) and its characteristics (excludability, rivalry, storage in the system).
- Identification of the characteristics of the communities (for example, social cohesion, homogeneity).
- Characteristics of the rules-in-use (formal, informal, enforced).
- Identification of the nature of the institutional arrangements present (centralized, decentralized, polycentric) (Alyward and Gonzalez 1998).
- Characteristics of the institutional arrangements (for example, economically efficient, equitable, accountable, adaptive) Alyward and Gonzalez 1998.
- Nature of feedback loops (for example, whether an actor providing a service receives something in return or not (reciprocity), or accountability/mandate of the mechanism).

This analysis would need to be carried out at three levels: for the providers of the watershed environmental services, for the receivers of these services, and for the existing or potential NSS. The overall analysis of a given situation based on the above parameters would provide insights into the kinds of incentives present at the levels of the WES provided, the community, and the rules-in-use towards watershed protection and management.

Analysis of External Actors

The second part of the framework helps analyse the impacts (intended and actualized) of the interventions of external actors on the incentive measures within a watershed system. This part consists of the following:

- Identification of the major external actors (such as the government or non-government organization [NGO]).
- Goals of the external actors.
- Identification of the nature of inputs provided (for example, technical, financial, knowledge, training).
- Nature of incentive measures (that is, market-based, regulation, information, cooperation) [OECD 1999].

While the first three parameters provide insights into the nature of the external actors and their interventions, from the point of view of WES, the critical parameter to examine here is the nature of incentive measures promoted by the external agencies. Each of the classifications suggested here (in the fourth point) can be further subdivided into specific measures to facilitate analysis. An indicative list for each is provided below:

- Market-based incentives: creation of markets/assignment of property rights/reform of adverse subsidies/subsidies that internalize externalities/compensation payments, fees, and taxes/ internal enforcement of contracts through revenue sharing.
- Regulation: Access restrictions and legal regulations of resource management.
- Information: Evaluation of non-market goods/publication and communication of valuation results/capacity building with regard to sustainable management techniques and external effects.
- Cooperation: Refers to participatory methods that ensure public acceptance, equity, and feasibility of institutional change.

The analysis of both parts of framework would help identify the key incentives, both within and outside the local community, for improving watershed management and the environmental services derived from it. It will also assist in analysing institutions at the community level in a manner that would provide insights into the nature of the negotiations utilized, and the sustainability of the mechanism. The framework would, therefore, serve as a common theoretical background on which to analyse or design processes for NSS for improved WES, their causes, and interventions.

APPLICATION OF THE ANALYTICAL FRAMEWORK TO THE BHOJ WETLANDS ACTION LEARNING PROJECT

Background

The Bhoj wetlands in the city of Bhopal, Madhya Pradesh, comprise two artificial lakes, namely, the Upper Lake and the Lower Lake. Constructed in the 11th century by King Bhoj of Dhar, the upper lake was created by building an earthen dam across the river Kolans. It has a catchment area of 361 km^2 and a water spread of 31 km^2. While its catchment areas are predominantly rural, there are some urban areas in the catchment as well.

The Bhoj wetlands are an important source of drinking water for the residents of Bhopal.[1] Additionally, both the lakes provide important recreational services such as boating and water sports

---

[1]For about 40 per cent of the city of Bhopal.

facilities. The wetlands also support a wide variety of flora and fauna. Over 160 species of birds and 14 rare macrophytes have been reported from the area. Considering its ecological importance, the Government of India declared the area as a Ramsar Site[2] in November 2002. The livelihoods of several communities are directly linked to the wetlands. A fishermen's cooperative consisting of some 500 fishermen's families has been given fishing rights on a long lease by the local authorities. Others cultivate water chestnuts in the wetlands, for sale locally. The wetlands have a high cultural value as well. The *Mazaar* (tomb of a Muslim saint) of Shah Ali Shah Rahamatulla Alliah, which has a religious significance, is located on Takia Island—a small island in the upper lake.

The Problem

Currently, the Bhoj wetlands is facing the twin problems of deteriorating water quality and reduction in storage capacity due to increased rates of pollution load and siltation. Water quality is being affected by a number of factors such as inflow of sewage and solid waste from the urban areas and run-off from agricultural farms in the peri-urban/ rural catchment of the upper lake.

There are approximately 87 villages in the upper catchment of the lake, falling within the districts of Bhopal and Sehore. Studies reveal that there are problems of nutrient (mainly nitrate and phosphate) and pesticide run-off and soil erosion[3] from the upper catchment into the lake, leading to pollution as well as siltation of the wetlands, affecting both water quality and storage capacity of the lake. The importance of addressing agriculture-related problems can be gauged from the fact that over 60 per cent of the catchment of the upper lake is cropland. Unless steps are taken to address the issue of agricultural run-off, the water quality as well as storage capacity of the Bhoj wetlands are likely to deteriorate further. This would affect the quality of drinking water supplied (and thus public health), as

---

[2]A *Ramsar Site* is a wetland that has been designated as internationally important according to a set of criteria under the terms of the (Ramsar) Convention on Wetlands, by the Administrative Authority of the Contracting Party State (*www.wetlands.org/RSDB/default.htm*).

[3]The major soil type in the catchment is 'black cotton soil' which is very prone to erosion.

well as the flora and fauna, livelihoods of fishermen, and the recreational value of the wetlands.

Exploring the Scope for Developing NSS

In this context, a change in management practices in the farmlands upstream could be a cost-effective and sustainable solution to the problem of agricultural run-off related pollution and siltation in the wetlands. This may be tackled by promoting wetland friendly agricultural practices, and building on the encouraging results obtained by pilot studies promoting organic compost as a replacement for chemical fertilizers under the Bhoj Wetlands Project.[4] This change in management practices is being piloted under a project implemented by Winrock International India (WII) in collaboration with the IIED, UK and the Lake Conservation Authority (LCA) in Bhopal. An attempt is being made to set up an NSS whereby the downstream receivers (city of Bhopal) provide incentives to the farmers in the upper catchment of the lake to promote the use of wetland-friendly agriculture practices, so as to reduce the siltation and pollution load in the lake.

An Institutional Analysis of the NSS

In this section the analytical framework suggested in the chapter is applied to the ongoing action learning project in the Bhoj wetlands. The institutional analysis of the efforts aimed at developing an NSS for WES from the rural catchment of the upper lake is described for 'local factors' and 'external factors'.

Local Factors

The WES in question in the case of the Bhoj wetlands are improved water quality and reduced silt load. The quality of the water that drains into the lake depends to a significant extent on agricultural practices within its rural catchment. Due to the prevalent use of fertilizers and pesticides and often inappropriate tillage practices, the chemical and silt loads of this water are high, which endanger the

---

[4]The 'Lakes Bhopal Conservation and Management Project', popularly called the Bhoj Wetland Project was implemented between April 1995 and June 2004. The project has played a significant role in reducing the pollution threat from urban areas, and some engineering structures and buffer zone vegetation belts have helped to control silt inflow from rural areas.

'health' of the Bhoj wetlands. Currently, mechanisms to control this inflow of pollutants and silt do not exist and chemical fertilizer usage patterns and detrimental agricultural practices are based on farming decisions taken by individual farmers.

The communities involved in the NSS process for the Bhoj wetlands include service providers and service receivers. The farming practices of most farmers in the catchment contribute to the pollution and silt loads flowing into the lake. However, in this case, the subset of farmers residing in eight villages in the riparian zone is considered as the potential service providers. Given their proximity to the lake or to major water channels that flow into the lake, their agricultural practices probably have the greatest impact on water quality.

There are several institutions that are the receivers of the services provided by the rural catchment of the upper lake. These include the Bhopal Municipal Corporation (BMC), industries, fishermen, laundry houses, water chestnut cultivators, hoteliers, tour operators, as well as the residents of Bhopal. For the purpose of this analysis, the BMC is considered as the primary service receiver since aspects of water quality and silt load have a major impact on its ability to supply sufficient clean water to consumers in Bhopal. If the water quality in the lake were to improve, BMC's treatment costs would drop and reduced silt loads would provide more water available for supply, thereby having a direct impact on the lives of at least 40 per cent of Bhopal's population. Further, BMC has jurisdiction over a portion of the riparian zone making it a stakeholder in the rural upper catchment as well.

However, the BMC is a complex institution driven by local political forces to a significant extent. Therefore, at this point, creating opportunities for productive face-to-face dialogues between the corporation and the riparian farmers is not an easy task. Consequently, other service receivers such as corporate and financial institutions as well as the tourism industry are being mobilized to work as supporting agencies towards the negotiation process and have been included as service receivers in the analysis.

Having defined the service providers and receivers, Table 3.1 provides an analysis of the current local factors as per the framework presented in Figure 3.1. Table 3.1 indicates that currently, in institutional terms, the service providers and the service receivers do not have any direct linkages. The BMC, which is considered as the primary service receiver, is an institutionalized body with its own formal set of rules-in-use that do not address the agricultural practices of farmers in the rural catchment of the lake.

The 'corporate/financial institutions/tourism industry' group of service receivers are profit oriented agencies driven by market forces with no direct linkages to the BMC or the riparian farmers. On the other hand, agriculture is the major source of income for riparian farmers who are potential service providers. However, these farmers are not 'organized' and do not have a set of common rules-in-use that could facilitate their active participation in negotiations for WES that they could provide.

Farming decisions of riparian farmers depend solely on the discretion and capacity of individual farmers and they are not accountable for the adverse impacts that these decisions have on the health of the lake. While the BMC is accountable to its consumers for the quality (and quantity) of water that it supplies in terms of appropriate treatment of contaminants, it does not have any institutional arrangement through which it can address the source of these pollutants. The 'corporate/financial institutions/ tourism industry' group of service receivers are accountable for their actions as per legislative directives. Therefore, not only are there no institutional linkages between the service providers and receivers, but even within the latter there are no such connections.

Given these local factors, in institutional terms, there does not exist any action situation (or common platform) where the actors (service providers and receivers in this case) can, or do, interact to develop an action arena within which an NSS (outcome as per the IAD) can spontaneously spawn. Further there are no rules-in-use that bind the various actors and influence or determine their behaviour. In fact, among the service providers and receivers there is little awareness of the potential of developing and implementing an NSS that could facilitate a change in agricultural practices, from inorganic to organic, which would have a direct bearing on the water quality and quantity of the upper lake.

Therefore, in order to bridge the gap between the service providers and receivers and to achieve the desired outcomes, an action situation will have to be developed within which interactions among the actors, based on defined rules-in-use, will have to be promoted. The efforts towards developing an NSS would, therefore, require a multi-pronged approach.

Table 3.1 Analysis of Local Factors and the Potential NSS

	Service Provider: Farmers in the riparian zone	Service Receivers: BMC (*Primary Receiver*)	Corporate/Financial Institutions and Tourism Industry	Potential NSS
Characteristics of the community	• Largely homogeneous (other backward classes) • Most landholdings greater than 10 acres • High dependence on agriculture for livelihoods	Regulatory service delivery body responsible for supplying water to 95 per cent of Bhopal's population of which 40 per cent receive water from the upper lake	Profit oriented	• Institutionalized farmer groups practising organic agriculture and working towards certification • BMC willing to provide incentives to riparian farmers to shift to organic farming • Corporate/financial institutions and the tourism industry willing to support the NSS process
Characteristics of the rules in use	Informal	Formal	Formal	Formalized rules-in-use between service provider and receivers
Nature of institutional arrangements	Polycentric	Centralized	Centralized	Intermediary body consisting of representatives from LCA, WII, accredited NGOs, and experts on organic farming constituted to facilitate transactions between riparian farmers and the BMC
Characteristics of institutional arrangements	Individual driven	Driven by political, policy, and legislative directives	Market driven	A combination of incentive-based and regulatory mechanisms that compensate farmers and help in monitoring practices and impacts
Nature of feedback loops	Not accountable	Accountable to consumers in Bhopal city	Driven by legislative directives	Reciprocal and accountable

Source: Authors.

- Mobilize riparian farmers to convert to organic farming and then organize them into groups which would facilitate the certification and the NSS process.
- Mobilize the BMC to recognize the positive role that such a conversion of farming practices would have on the quality and quantity of water in the upper lake and generate an institutional willingness to support this conversion.
- Mobilize corporate and financial institutions and the tourism industry to support the negotiation process between the riparian farmers and the BMC, and thus contribute to the bundle of incentives provided to the service providers.

The proposed intermediary body, consisting of experts and practitioners, which would coordinate the NSS is aimed at developing an action situation, facilitating interactions among the actors, and defining the rules-in-use. This body is expected to create an appropriate institutional environment which will facilitate the development of institutional arrangements that are acceptable to the service providers as well as the service receivers. This intermediary body would 'absorb' potential conflict situations and other differences that may hamper the transaction process. It would facilitate the smooth functioning of the transaction system and facilitate the provision of technical inputs to the farmers. This body would also play a monitorial role to ensure that reciprocity and accountability are maintained within the NSS (feedback loops).

External Factors

Since developing NSS for protecting and managing WES is a new and untested method (in India), supportive external factors are critical to introduce and facilitate this process. The major external actors in this case are WII (including IIED support) and the LCA of Madhya Pradesh. Table 3.2 provides an outline of the nature of these external factors.

From Tables 3.1 and 3.2 it is apparent that the process of developing NSS in this case is contingent on the ability of external actors to mobilize stakeholders, and create an awareness regarding the problem and acceptance of the proposed solution. Therefore, clearly the 'push factor' is greater than the 'pull factor'. This is expected since the 'proposed solution' calls for radical changes in land use practices and mindsets. Conversion of agricultural practices from inorganic to organic has associated losses in the first two years that

Table 3.2 Nature of External Factors

	Details/Characteristics
Major external actors	Winrock International India/Lake Conservation Authority
Goals of external actors	To develop incentive-based mechanisms to protect and manage WES provided by the rural catchment of the Bhoj wetlands
Nature of inputs provided	• Mobilizing service providers and receivers and other supporting agencies (corporate or financial institutions) • Commissioning studies (socio-economic/technical/policy level) to create baseline and supporting information • Facilitation of dialogue between relevant stakeholders • Training and awareness generation programmes • Creating institutionalized farmer groups practising organic agriculture • Access to technical support (for conversion to organic farming/certification) • Developing and promoting an incentive-based mechanism for WES and improved livelihoods • Establishing an 'intermediary body' that will facilitate and coordinate the negotiation support system
Nature of incentive measures	The nature of the incentives are a combination of: • Information: through training, access to knowledge and awareness generation through publications • Cooperation: through mobilization of providers and receivers and other supporting agencies and creating greater awareness amongst the public • Market-based Incentives: creating internal control systems within farmer groups to facilitate the organic agriculture certification process/potentially lower water treatment costs for the BMC

Source: Authors.

the farmers have to absorb, after which agricultural productivity would increase and product value would rise to levels above those achieved through inorganic farming. The NSS does not support providing monetary compensation to farmers during this 'loss making phase' but instead chooses to pay for providing farmers with access to technical know-how and the certification process. These extension services are to be coordinated by the intermediary body. This intermediary body, which is critical to the success of the NSS, would require financial assistance for functioning and for hiring the services of experts who can guide and train the farmers in the conversion to organic farming. These finances are being sought from corporate organizations such as HEG Ltd under their Corporate Social Responsibility Programme and from financial institutions such as the National Bank for Agriculture and Rural Development (NABARD) and the State Bank of India (SBI) at discounted rates of interest. This financial support to the intermediary body is critical since it is expected that getting BMC actively involved in the NSS would not be an easy task. It is hoped that initiating the process and seeing some tangible results would mobilize the BMC to partner with the NSS process.

The incentives are anticipated to be based on measures of information, cooperation, and market-based incentives, as mentioned in Table 3.2. While information and cooperation are supportive incentives, the market-based incentive is in a sense an internal regulatory one that ensures that agricultural practices within a farmer group remain based on the principles and practices of organic farming and facilitates certification.

Based on the nature of inputs provided by the external actors (WII/LCA), it seems that their efforts are at aimed defining and developing an action arena within which the NSS can be situated. The establishment of the intermediary body is an attempt to create an action situation (as mentioned earlier), while the mobilization of the service providers and receivers—through awareness generation activities, facilitation of dialogue, and training programmes is aimed at influencing the behaviour of the actors and defining mutually acceptable rules-in-use. Interactions between the actors, that is, between riparian farmers and BMC with support from corporate/financial institutions and the tourism industry, are to be channelled through the intermediary body. Thus, this body would play a critical role in defining the action arena, and consequently would significantly impact the institutional arrangements within the NSS for the Bhoj wetlands.

The NSS

The analysis of the local and external factors that affect the development of NSS for the protection and management of WES in the Bhoj wetlands helps identify the key incentives, the nature of the negotiations, and the sustainability of the mechanism.

KEY INCENTIVES

1. For the riparian farmers (service providers)
   - Improved agricultural productivity through organic farming (after a period of two years)
   - Certification of their agricultural products and thereby access to niche markets that would fetch higher prices—greater income per unit of land
   - Monetary savings on agricultural inputs
   - Access to technical inputs and capacity building opportunities related to organic farming and improved agricultural practices at no cost
   - Sustainable utilization of land resources that will lead to sustainable livelihoods in the long term
2. For BMC and corporate/financial institutions and tourism industry (service receivers)
   - Improved water quality and quantity in the upper lake
   - Lower costs of water treatment
   - Storage capacity of the upper lake improved and therefore the ability to supply water to the residents of Bhopal, especially during the dry season
   - 'Health' of the Bhoj wetlands maintained
     - Unique selling proposition (USP) for the tourism industry (increased recreational value of the lake)
     - Tangible contributions of the corporate sector under their corporate social responsibility programmes

NATURE OF THE NEGOTIATIONS

The NSS would have to be able to clearly describe, based on validated facts, the win-win situation that it aims to achieve to the riparian farmers as well as the BMC, corporate/financial institutions, and the tourism industry and generate a willingness among these actors to participate in the process in a meaningful manner. Further, it would have to strongly promote interactions (through dialogue) between these actors. Based on these interactions, the NSS process would need to define the institutional environment (rules of the game) and mutually acceptable institutional arrangements (rules-in-use) that would be the most important independent variable driving the process.

Further, the nature of the negotiations would need to be integrative and dynamic so that they are able to identify new problems and address wider issues that primarily affect riparian farmers (service providers). This is especially important as currently riparian farmers are not accountable for their agricultural practices and because there is a loss associated with conversion to organic farming in the first two years—both of which could work as disincentives for farmers to shift to organic farming. An NSS that would address other issues affecting their lives and livelihood would probably engender greater willingness amongst the riparian farmers to radically change their agricultural practices.

SUSTAINABILITY OF THE MECHANISM

The sustainability of the NSS hinges on three critical parameters. First, and foremost, the sustainability of the mechanism would depend on the willingness of the service providers and service receivers to actively participate in the mechanism and play out their expected roles. Second, the system's sustainability would be determined by the manner in which the intermediary body is able to facilitate and coordinate the negotiation system and provide critical inputs to support it. Presuming that this institution has the capacity to function effectively, it will require financial and state policy support to implement its responsibilities. Finally, the sustainability of the mechanism would be contingent on the actualization of the adoption of organic farming which is one of the key desired results of this effort at developing an NSS for the protection and management of WES in the Bhoj wetlands.

SUMMARY

In the case of the Bhoj wetlands, external factors play a pivotal role in the development of a successful NSS. Though a strong case for developing such a system can be justified, the current institutional environment between service providers and receivers is not conducive and would require the intervention of a third intermediary party. Therefore, effectively, the push factor applied by external actors would need to generate and cultivate an associated pull factor from the side of the local actors. This would need to be done through extensive mobilization, awareness generation, and capacity building campaigns across all actors, as well as the institutional development of the service providers who are currently disaggregated as far as developing a workable NSS is concerned.

Further, efforts at developing such a negotiation support system will need to focus on developing the following:

1. an arena where the concerned actors would be willing, and able to, interact (establishment and effective functioning of the intermediary body);
2. mutually acceptable institutional arrangements (rules) that these actors would be willing to adhere to and accordingly modify their behavioural patterns (service providers: conversion to organic farming; service receivers: support service providers to convert to organic farming); and
3. generate the support of policy makers and political actors towards the NSS proces.

CONCLUSION

As stated earlier and demonstrated through the application of the analytical framework on an ongoing initiative in the Bhoj Wetlands, managing and maintaining WES represents a complex collective action problem. Efforts towards developing NSS between the providers and receivers of these services have to tackle a multi-tiered institutional maze, especially when there are several actors involved, as is the case with the Bhoj Wetlands initiative. The framework presented in this chapter represents an analytical institutional tool that facilitates the deciphering of this maze. It can be applied to develop a 'guide map' to be followed, or to assess the accuracy and effectiveness of the 'route' taken.

By disaggregating local and external factors, the framework allows for an assessment, or matching, of inputs provided by external agencies to the situation and requirements at the local level. The most critical is a thorough analysis of the local factors that would influence the development of an NSS. An NSS for WES necessarily requires a social space, or action situation, within which providers and receivers can interact. As found in the case of the action learning project in the Bhoj Wetlands, this space does not naturally exist and attempts are underway by external actors to establish an intermediary body that would help create such a space.

The development of the NSS is the desired outcome of an intervention, such as the ongoing one in the Bhoj Wetlands. As per the fundamental structure of the framework (which corresponds to that of the IAD framework and has already been discussed), the outcomes are produced within the action arena, through activities, interactions, and exchanges among the actors. The action arena is, therefore, defined by the actors and the action situation, and is influenced by the characteristics of the physical world (WES in this case), the communities involved (service providers and receivers), and the rules-in-use. Therefore, in the absence of a conducive action situation, defining the action arena would not be possible. Further, mutually acceptable rules-in-use would be developed through interactions between actors within an action arena.

Therefore, in cases such as the Bhoj Wetlands initiative, the external actors have to critically focus not only on mobilizing and empowering local actors (through training, awareness generation, and capacity building), but on developing a social space for the latter to interact in as well. In cases where there is a single service provider and one corresponding receiver (the providers and receivers can be a group of individuals represented by a single institution) there are chances that facilitating a change in behaviour and resource use practices can lead to the development of a social space for the NSS to develop in. However, with numerous actors, as in the Bhoj Wetlands, most often, specific efforts towards creating this social space would have to be allocated. Either way, it is within this social space that actors would come together and interact for the development of an incentive-based, mutually acceptable, and sustainable NSS for managing and maintaining WES.

REFERENCES

Alyward, B. and Alvaro Fernandez Gonzalez (1998), 'Institutional Arrangements for Watershed Management: A Case Study of Arenal, Costa Rica', Working Paper No 21, Collaborative Research in the Economics of Environment and Development (CREED), London.

Bates, R.H. (1988), 'Contra Contractarianism: Some Reflections on the New Institutionalism', *Politics and Society*, Vol. 16, Nos 2 and 3, pp. 387–401.

Bravo, G. (2002), 'Environment Institutions and Society in the Management of Common-pool Resources: Linking IAD Framework with the Concept of Social Capital', Presented at the Ninth Conference of the International Association for the Study of Common Property, Victoria Falls, Zimbabwe, 17–21 June.

Fischer, A. and Lorenz Petersen (2004), 'How Incentives Matter—A Conceptual Framework for Natural Resource Governance in German Development Cooperation', Presented at the Conference on International Agricultural Research and Development, Berlin, Germany, 5–7 October, 2004.

Gibson, Clark C. (2005), 'In Pursuit of Better Policy Outcomes', *Journal of Economic Behaviour and Organisation*, Vol. 57, pp. 227–30.

Hardin, G. (1968), 'The Tragedy of the Commons', *Science*, Vol. 162, pp. 1243–8.

Mulholland, S. and Christine Shakespeare (2005), 'An Analysis of Frameworks for Higher Education Policy Studies', AIHEPS, New York University.

OECD (1999), *Handbook of Incentive Measures for Biodiversity—Design and Implementation*, OECD Publications, Paris.

Ostrom, E. (1990), *Governing the Commons: The Evolution of Institutions for Collective Action*, Cambridge University Press, Cambridge.

Ostrom, E., R. Gardner, and J. Walker, (1994), *Rules, games and Common Pool Resources*, Michigan University Press, Ann Arbor.

Thomson, J. (1992), *A Framework for Analysing Institutional Incentives in Community Forestry*, FAO, Rome.

Thomson, J. and K. Schoonmaker Freudenberger (1997), 'Crafting Institutional Arrangements for Community Forestry', *FAO Community Forestry Field Manual*, FAO, Rome.

4

Incentives that Work for Farmers and Wetlands
Analysis from a Choice Experiment at the Bhoj Wetlands, India*

ROBERT A. HOPE, MAMTA BORGOYARY, AND CHETAN AGARWAL

INTRODUCTION

The Millennium Ecosystem Assessment Synthesis on Wetlands and Water recognizes that modern agricultural systems have delivered significant economic and food benefits at the cost of being a high impact and rapid driver of global wetland degradation (MA 2005). Many governments are now considering alternative approaches in response to increasing awareness about degrading water resources and growing concern over the significant fiscal burden of agricultural subsidies (World Bank 2003). Organic farming represents one alternative approach which is gaining public and policy support due to a range of demonstrated benefits, including supporting higher levels of biodiversity, conserving soil fertility and stability, reducing energy demand, and improving ground and surface water quality from non-use of herbicides and pesticides[1] (Stolze et al. 2000; Environment

*This study benefited from the generous support of Pradeep Nandi and his colleagues at the LCA of Madhya Pradesh. Many useful comments and advice were given by Vivek Sharma and his team at the Centre for Advanced Research and Development (CARD), Bhopal, which managed the fieldwork. This study complements a wider international study led by the IIED and thanks are extended to Elaine Morrison, Ivan Bond, and Ina Porras for supporting this collaborative effort. This publication is an output from a research project funded by the United Kingdom Department for International Development (DFID) for the benefit of developing countries. The views expressed are not necessarily those of DFID (R8174–Forestry Research Programme).

[1] It is noted that switching to organic inputs does not necessarily reduce problems of excess nutrient loads in water bodies (eutrophication) but non-use of herbicides

Agency 2002). However, a common barrier to farmers shifting to organic land management is a likely short-term reduction in food production and income, which represents a significant adoption constraint for small-scale farmers in tropical countries (Harriss et al. 2001; Ghosh 2004).

Innovative financial mechanisms, such as PES, provide an emerging framework to facilitate small-scale farmers' decisions to adopt more benign land practices by making more explicit relationships between upstream environmental service providers (for example, farmers, pastoralists, and forest dwellers) and downstream environmental service users (for example, nature, industry, and water supply). In theory, people providing environmental services are compensated to maintain an agreed level of service provision contingent on users receiving measurable and, preferably additional, service benefits (Wunder 2005). In relation to watershed services, such as maintaining water quality, it can be conceived of as a negotiation process in which upstream land managers agree to an opportunity cost of modifying land use behaviour that is paid for by downstream users' associated benefits or cost avoidance. There is widespread interest and policy support to better understand the opportunities from incentive mechanisms to improve environmental management and contribute to rural development (van Noordwijk et al. 2004; Wunder 2005). Pagiola et al. (2004) note that a mechanism may not require payments in perpetuity (with associated transaction costs and risks of non-compliance) but may be effective as a transitional or bridging payment if new practices are more profitable to land users than existing practices (for example, see Bassi 2002).

In Asia, wetlands are considered severely under threat by aspects of rapid economic growth (for example, industry, urban expansion), including off-site water quality damage from agricultural pollution (IUCN n.d.). In India, Ghosh (2004) argues that a shift to organic farming is urgently needed on ecological grounds and provides evidence that shifting to the same is feasible for farmers in almost all states on financial criteria. A country-level analysis by the World Bank (2003) argues that 'rapidly rising (agricultural input) subsidies are fiscally unsustainable and crowding out productivity-enhancing public investments in rural infrastructure, irrigation, and technology upgrading'. However, policy change is challenged by the scale and

---

and pesticides is associated with on-site and off-site water quality improvements, and will reduce inorganic fertilizer costs associated with energy consumption and fiscal subsidies.

depth of poverty in India, which is home to one-third of the world's poor people, who mainly live in rural areas and are largely dependent on agricultural livelihood activities (World Bank 2003). Demand for organic food in developing countries such as India, is growing though small when compared with the growing international demand for organic produce estimated at US$ 11 billion, of which imports from developing countries accounted for US$ 500 million (Harriss et al. 2001). Small-scale farmers from developing countries face a number of significant constraints accessing national and international markets. In particular, certification of organic produce is often an absolute requirement (for example, European Union) and, in the case of small-scale farmers, organization into producer groups is essential for cost-effective group certification. In a global review, Harriss and colleagues conclude that despite cost, information, and scale constraints, 'there is evidence that resource-poor smallholder farmers can obtain economic and social benefits from participation in organic production and trade' (Harriss et al. 2001, 51). As such, a shift to organic farm management offers one approach that may partly respond to economic growth, poverty reduction, and wetland conservation goals.

Understanding which incentives will influence small-scale farmers to commit to switching to organic farming practices is key to designing effective and sustained policy action that works for farmers and the environment. This chapter presents research from the Bhoj wetland in India, which explores the feasibility of farmers adopting organic farming as a measure to reduce pollution run-off into an ecologically-significant wetland site. A stated choice experiment analyses farmer preferences to organic farm management scenarios to provide guidelines for the incentives and policy action that are most likely to influence a shift to organic farm management. The rest of the chapter is organized as follows. The next section introduces the study location; the third section describes the methodology; the fourth section presents the study results, and the fifth section concludes with policy implications.

STUDY LOCATION

Madhya Pradesh has one of the highest concentrations of rural poverty in India with approximately one in three people classified as poor (Deaton and Dreze 2002[2]). Poverty incidence is highest

---

[2]The authors debate and estimate national estimates of poverty based on three national sample surveys.

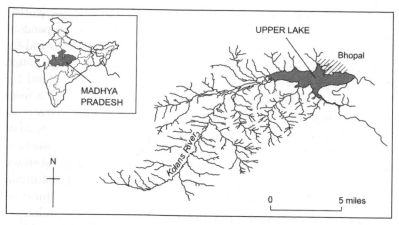

Map 4.1:   Location of Villages in Kolans Watershed

amongst small-scale farmers with scheduled castes (SC) and scheduled tribes (ST) more likely to be poor than other social groups (World Bank 2003). Agriculture represents the main livelihood of the state's 60 million people. Farmers commonly use fertilizers and pesticides for a monsoon crop (59 per cent and 28 per cent of farmers, respectively) and a post-monsoon crop (52 per cent and 23 per cent of farmers, respectively[3] (GoI 2005). Increasing off-site water quality impacts from rural agricultural land use are a concern for the Government of Madhya Pradesh (Verma 2001). Impacts at the Bhoj wetland are of particular concern as it is the Ramsar site of international ecological significance and provides 40 per cent of the drinking water supply to the state capital, Bhopal (Figure 4.1).

The Bhoj wetland is located in the Kolans watershed and dates back to the 11th century when Raja Bhoj of Dhar built an earthen dam across the Kolans river. The wetland constitutes an upper and lower lake; the upper lake is the major water body. The upper lake measures 14 km in length and varies between 2 km and 12 km in width, covering an area of 36 km^2. The average lake depth is 4 metres with the deepest point reaching 14 metres. The upper lake is classified as mesotrophic with pollution sources derived from urban and rural sources (Mishra 2006). Situated in the Vindhyan range on the borders of the Malwa plateau, the main geological formations are Bhander sandstone and Deccan trap lava flows. These contribute to good

---

[3]The monsoon crop is called 'kharif', a post-monsoon crop is called 'rabi', a third crop, after the rabi, may be also be grown called the 'zaid'.

black cotton soils and with a gentle topography and average rainfall greater than 1200 mm, agriculture is the main land use across the 87 villages in the peri-urban and rural areas of the watershed (Verma 2001). Census data from 2001 estimate 14,109 households with 83,909 people located in the watershed.

The wetland provides important cultural, water supply, and environmental services. The wetland supports a wide variety of flora and fauna, including several species of phytoplankton and zooplankton, aquatic insects, amphibians, fishes, and birds (resident as well as migratory) (Borgoyary 2005). The upper lake of the wetland provides 40 per cent of the drinking water supply to the 1.8 million residents of Bhopal (Verma 2001). Livelihoods of many people are also directly linked to the wetland through fishing, farming and recreation. The wetland has important socio-cultural values represented by the location of the mazaar (tomb of a Muslim saint) of Shah Ali Shah Ali Shah Rahamatulla Alliah on Takia Island.

Urban pollution is linked to the growth of Bhopal, which has developed rapidly on the borders of the wetland over the past 50 years. Urban pollution includes various industrial effluents, idol immersion, laundry houses (*dhobi ghats*), human sewage, and chemical inputs for water chestnut farming. A Japan Bank for International Cooperation (JBIC) project in the 1990s helped address many of the urban pollution issues, in partnership with the Government of Madhya Pradesh. Interventions included buffer zones between the lake and the city (forestry and roads), building over 85 km of new sewage pipes to divert 56 million litres of sewage per day, re-locating dhobi ghats away from the main lake and collaborating with the Government of Madhya Pradesh to set up the LCA, which acts as a state-wide resource for scientific research and policy on management of the state's water bodies (Borgoyary 2005).

Since the 'Grow More Food' campaign to tackle food shortages in the 1970s, there has been widespread use of fertilizers, herbicides, and pesticides in the watershed, which has contributed to the degradation of on-site soil and off-site water resources (Mishra 2006). Since 2004, the LCA has piloted extension activities to demonstrate organic farming techniques, such as vermi-composting and improved composting of farm yard manure, to some peri-urban farming communities but uptake has been slow and uneven (P. Nandi, personal communication 2005). Without improved understanding of farmer preferences to incentives to shift to organic farm management, decision makers have insufficient information to evaluate the value and impact of wider promotion of organic farming across the watershed against socially-acceptable and wetland-wise use criteria.

METHODOLOGY

Choice experiments

Choice experiments provide an approach to estimate stakeholder preferences to a predicted or planned future scenario that cannot be objectively assessed with existing knowledge (for example, climate change, price shifts, new technology). It allows decision makers to explore behavioural responses and priorities, which cannot be satisfactorily answered by analysis of observed data. A feature of the methodology is that a good or service is characterized by a collection of attributes and attribute levels (Lancaster 1966) rather than assuming an aggregate 'whole good' approach. As such, experimental design presents choice tasks in a trade-off or voting game format, which allows assessment of the marginal value respondents place on each attribute. The methodological framework is well-known and is discussed extensively elsewhere (see Louviere et al. 2000).

A limitation of the method is that choices are often shaped by the way they are framed. Scoping analysis and identification of attributes are critical to informing a valid and legitimate experimental design. Detailed scoping work with institutional actors and stakeholder groups was conducted to inform a pilot choice design (Borgoyary, 2005). During a training workshop in the pilot phase, a locally-based NGO (Centre for Advanced Research and Development) scrutinized the design with other interested institutional actors (including LCA), which attended field-testing of three pilot designs in watershed communities (Hope et al. 2005). During this process, the final questionnaire and choice experiment design were agreed upon.

Study design

The household questionnaire captured data related to current farming practices and existing organic farming practices. The final version of the questionnaire was translated into Hindi and reviewed by a bilingual Hindi–English member of the research team to test for any inconsistencies or anomalies in language, sense, or interpretation. The questionnaire has four sections: (i) household selection and data

Table 4.1: Sample frame

ID	Zone	Village	Sample size (farmer households)	No. of Sets (units of 8)
7	BMC	Barkheda Nathu	48	6
28	BMC	Goria	8	1
58	BMC	Malikhedi	24	3
61	BMC	Mugaliyachhap	56	7
66	BMC	Neelbad	8	1
		sub-total	144	18
32	LOWK	Int Khedichhap	32	4
37	LOWK	Kajlas	32	4
41	LOWK	Khajoori Sadak	64	8
52	LOWK	Kolu Khedi	32	4
56	LOWK	Lakhapur	48	6
70	LOWK	Pipaliya Dhakad	40	5
		sub-total	248	31
16	UPK	Bilkisganj	48	6
23	UPK	Dhabla	48	6
43	UPK	Kharpa	8	1
44	UPK	Kharpi	48	6
55	UPK	Kulas Khurd	48	6
87	UPK	Uljhawan	48	6
		sub-total	248	31
		Total	*640*	*80*

Source: Authors.

quality; (ii) farming system; (iii) choice experiment; and (iv) household characteristics (Hope et al. 2005). The sampling frame identified three sampling zones across a representative group of 17 communities in the Kolans watershed: (i) Bhopal Municipal Corporation (BMC)—villages located in a peri-urban riparian zone; (ii) lower kolans watershed (LOWK)—villages located in the lower but more rural riparian watershed zone; (iii) upper kolans watershed (UPK)—villages located in the more remote upper watershed zone (Table 4.1). Sampling within villages was randomized within the purposive constraints indicated here.

Experimental design

Following the pilot phase, attributes and attribute levels that best responded to the research question were chosen (Table 4.2). All attributes were presented pictorially to increase the ease with which many illiterate farmers could participate in the study. No farmer

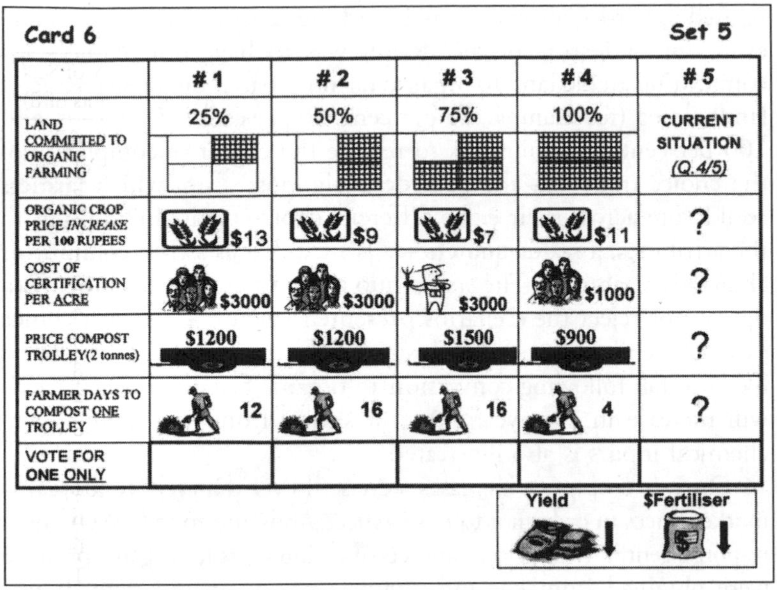

Fig. 4.1   Choice card example

Table 4.2:   Choice attributes and attribute levels

Attributes	Levels					
Land commitment to organic farming (acres)	25%	50%		75%		100%
Organic crop price increase per 100 Rupees	5	7	9	11	13	15
Cost of certification per acre	Rs 1000 as a group		Rs 3000 as a group		Rs 3000 as an individual	
Compost price per trolley (Rupees)	Rs 600		Rs 900	Rs 1200	Rs 1500	
Days to compost per trolley	4		8		12	16

Source: Authors.

refused to participate and many found the experiment intriguing and even fun. A feature of the design was to have four choices for committing arable land to organic farming based on a percentage of landholding (for example, 25 per cent, 50 per cent, 75 per cent, and 100 per cent). The aim was to reduce the cognitive complexity of the choice trade-offs and provide a 'signpost' from which farmers could vote across their eight different choice cards. In addition to the attributes, a status quo choice is included as a fifth column (or choice) in each card. The status quo option permits respondents to opt out or reject the scenarios presented (Figure 4.2). Each choice card also reminds farmers with simple illustrations that crop yield is likely to fall following conversion to organic farming though yields will increase in later years. A cost saving from not buying agro-chemical inputs is also illustrated.

Organic crop price increases were set lower than reported organic market prices in order not to raise expectations and to reduce potential response centrality to price. Land certification costs for organic produce were obtained from key informants working in the state. It was expected that, all things being equal, farmers would prefer lower to higher certification costs, therefore, it was considered sensible to include a lower cost group attribute with two higher but same cost attributes that were distinguished by working as a group or individually. This aimed to untangle 'cost' and 'collective' signals from the data. Current local compost[4] prices were estimated from existing local prices. Labour days to make one trolley of compost were estimated and an appropriate range was chosen (Borgoyary 2005).

The attribute levels result in a 4^3*6*3 factorial design with 16 main effects and 100 two-way interactions. A mains effects orthogonal design function was run in SPPS (version 11.5) resulting in a 64 card design with 8 cards repeated. Eliminating duplicate cards reduces orthogonality and duplicate cards were left in the design (Louviere et al. 2000). It was decided that respondents were able and willing to answer up to 8 choice cards each. This required the questionnaire to be rotated in units of eight (For example, Set 1 through to Set 8 are administered to the first eight respondents with

[4]Manure is a collected heap of farm yard waste which may have decomposed but in an uneven and untreated manner. Manure and the term Farm Yard Manure are used interchangeably here. After some level of manure treatment and decomposition the output would be of higher agricultural value and is here termed compost.

respondent nine starting with Set 1 again, and so on). To confirm that respondents had the same information about the experiment, an introductory text was explained to every respondent, including details on: (i) impacts of chemical agriculture; (ii) shifting to organic farming; (iii) the voting game; (iv) voluntary and serious exercise; and (v) testing a dummy card. The dummy card allowed enumerators to test that each respondent had understood the experiment and provide a 'dry run' to clarify any queries or misunderstandings before the eight 'live' choice cards were voted on.

RESULTS

Descriptive data analysis

Descriptive data analysis of farm profile by location is presented in Table 4.3. Results for household characteristics, assets, and farm management are decomposed by watershed zone. Household composition indicates that families consist of around seven people of Hindu faith belonging to SC, ST, or other backward castes (OBCs). Heads of household are almost all male with one in three illiterate and generally under 50 years of age (57 per cent). Homes are mainly classified as 'poor' (kaccha) with sanitation access improving from seven out of ten households in the upper kolans (UPK) using an open field to under six out of ten households using

Table 4.3: Farm profile by location

	UPK (n=248)	LOWK (n=248)	BMC (n=144)	ALL (n=640)
Household characteristics@				
Household size	6.44 (0.11)	6.77 (0.12)	7.16 (0.14)	6.79 (0.07)
Head illiterate	0.36 (0.02)	0.37 (0.02)	0.36 (0.02)	0.36 (0.01)
Head over 50 years	0.42 (0.02)	0.40 (0.02)	0.46 (0.02)	0.43 (0.01)
SC/ST/OBC	0.85 (0.01)	0.81 (0.02)	0.94 (0.01)	0.87 (0.01)
Hindu	0.97 (0.01)	0.81 (0.02)	0.94 (0.01)	0.87 (0.01)
Poor (kaccha) home condition	0.55 (0.02)	0.63 (0.02)	0.56 (0.02)	0.58 (0.01)
Open field sanitation	0.70 (0.02)	0.66 (0.02)	0.58 (0.02)	0.65 (0.01)
Drinking water > 500 metres	0.25 (0.02)	0.43 (0.02)	0.27 (0.02)	0.32 (0.01)

(contd...)

Table 4.3 (contd...)

	UPK (n=248)	LOWK (n=248)	BMC (n=144)	ALL (n=640)
Assets#				
Annual income (US$)	1335 (117)	1621 (85)	1879 (93)	1612 (57)
Cultivation as a proportion of annual income	0.73 (0.01)	0.76 (0.01)	0.67 (0.01)	0.72 (0.01)
Income poor household*	0.80 (0.02)	0.68 (0.02)	0.62 (0.02)	0.70 (0.01)
Land owned (acres)**	9.33 (0.41)	8.12 (0.36)	9.46 (0.51)	8.97 (0.25)
Tractor	0.22 (0.02)	0.34 (0.02)	0.37 (0.02)	0.31 (0.01)
Bullock cart	0.51 (0.02)	0.26 (0.02)	0.32 (0.02)	0.36 (0.01)
Water pump	0.44 (0.02)	0.51 (0.02)	0.49 (0.02)	0.48 (0.01)
Farm management$				
Land cultivated (acres)*	9.91 (0.43)	9.13 (0.43)	12.10 (0.67)	10.37 (0.30)
Land farm organically (acres)**	0.83 (0.05)	0.56 (0.03)	0.69 (0.04)	0.69 (0.02)
Percentage of farm ... for fuel cakes	42.29 (0.89)	46.29 (0.88)	41.40 (0.93)	43.25 (0.52)
dung ... ... for manure	54.02 (0.94)	51.99 (0.89)	54.15 (0.98)	53.42 (0.54)
Kharif crop sold (as per cent of harvest)	79.54 (0.96)	80.26 (0.98)	73.39 (1.07)	77.73 (0.58)
Rabi crop sold (as per cent of harvest)	53.02 (1.12)	57.36 (1.07)	49.41 (1.18)	53.26 (0.65)
Fertilizer use***(kg per acre per year)	131.90 (3.70)	157.41 (3.39)	136.21 (3.95)	141.79 (2.14)
Herbicide and pesticide use****(litres per acre per year)	0.42 (0.02)	0.54 (0.02)	0.39 (0.02)	0.45 (0.01)

Notes: All data reported as mean (standard error); zones weighted by inverse probability of selection.

@Household characteristics—all but household size are dummy variables (1 = variable name; 0 = head literate, head under 50 years, other caste group, Muslim, semi-kaccha or pucca home condition, pit latrine or better, drinking water <500 metres, respectively);

#Assets—*income poverty is estimated from an adult equivalent scale [1+0.7*(adults over 16 years -1) + 0.5*(Children under 16 years)] with less than US$1 per adult equivalent per day indicating an income poor household; ** land owned indicates 'paper' proof of ownership and excludes land leased-in or communal land use (see below); tractor, bullock cart and water pump are dummy variables as above;

$Farm management—*includes leased-in land and land with no papers (for example, communal); **average acres of land farmed organically, including farm yard manure, compost, vermi-compost; ***outliers (>350 kg per acre) are excluded (n=579); ****outliers (>2.0 litres per acre) are excluded (n=584).

Source: Authors.

an open field in the BMC. One in four households in the UPK and BMC access water from over 500 metres away, though this falls to four in ten households in the LOWK.

A combination of the opportunity cost of land (for example, urban expansion) and increasing access to non-farm employment may largely explain increasing income and falling dependency on cultivation as household location moves from UPK to LOWK to BMC. There is an expected rise in annual income with increasing proximity to the city of Bhopal as described by a household income of US$ 1335 in the UPK compared to US$ 1879 in the BMC. This is associated with a reduction in cultivation as a proportion of income from 0.73 to 0.67, respectively. This is also reflected by a fall in the incidence of a household being defined as income poor (here, US$ 1 per person per day) from 0.80 in the UPK to 0.68 in the LOWK to 0.62 in the BMC.

On average, households own nine acres of land. However, many households lease-in or farm communal land to increase average area cultivated to over ten acres per farm. Of the cultivated land, less than one acre is farmed organically as defined by using compost, farm yard manure, or vermi-compost. Households in the UPK currently farm the largest organic area (0.83 acres) with slightly lower areas reported by households in the LOWK (0.56 acres) and BMC (0.69). It is noteworthy that household allocation of farm manure is split relatively evenly between compost (53 per cent) and fuel cakes (43 per cent). This indicates that, given an acceptable energy substitute, existing farm manure could be used to increase organic production for the average household. Household allocation of crops' harvest by season are consistent with wider practices of selling a higher proportion of kharif (monsoon) crops (roughly, 80 per cent of total) than rabi (non-monsoon) crops (roughly, 50 per cent of total). Tractor-ownership ranges from roughly one in five households in the UPK to over one in three households in the LOWK and BMC. Tractor-ownership is inversely mirrored by greater use of bullock carts in the UPK (one in two households) compared to less than one in three households with a cart in the other two zones. Water pumps are reported by almost half of all respondents across the watershed.

Chemical-based agricultural inputs are estimated for application to an average acre per year for fertilizer use (kg) and a combined estimate for herbicide and pesticide use (litres). Fertilizer use is reported at 140 kg per acre per year with herbicide and pesticide use at 0.45

litres per acre per year for an average farm. Given the Kolans watershed measures 361 sq. km and the majority of land is cultivated, input use can be roughly calculated based on a 'back-of-envelope' estimate. For example, if 50 per cent of the 14,000 households farm an average 10 acres, this represents 70,000 acres or 80 per cent of total watershed area. This would tentatively indicate that 9800 tons of fertilizer and 31,500 litres of herbicide and pesticide are applied in the watershed per year. While shifting to organic farming will not necessarily reduce nitrate concentrations in water bodies without other land management interventions (for example, zero tillage, bunding, riparian protection), it would require no further farm use of herbicides or pesticides in order to qualify for organic land certification.

Choice experiment analysis

Each of the 640 respondents successfully completed their eight choice tasks in the questionnaire with only four errors in coding the choice data providing 5116 observations for analysis. Analysis of choice responses is based on random utility theory which models the probability that individual n will choose option i over any other option j belonging to the complete choice set C as:

$$P_{in} = p[(V_i + e_i) > (V_j + e_i), \forall_j \in C] \qquad \ldots (4.1)$$

where V is an observable systematic component of alternative i and e is a random component. A typical assumption is that the random component is independently and identically distributed (IID) with an extreme value distribution (Hausman and McFadden 1984). Under this assumption, an explicit form of the probability of choice of Equation (4.1) is satisfied by the multinomial logit model (McFadden 1973):

$$P_{in} = \exp(\lambda V_{in}) / \sum_j \exp(\lambda V_{jn}), \forall_j \in C \qquad \ldots (4.2)$$

where[a] is a scale parameter which is inversely proportional to the standard deviation of the error terms and V_{in} and V_{jn} are conditional indirect utility functions assumed to be linear in parameters,

$$V_{jn} = C_j + \sum \beta_{jk} X_{jk} + \sum \gamma_{jn}(S_n {}^* C_j) \qquad \ldots (4.3)$$

where C_j is an alternative specific constant, X_{jk} is the k attribute value of alternative j, β_{jk} is the coefficient associated to the k attribute,

S_n is the socio-economic characteristics' vector of individual n and γ_{jn} is the vector of the coefficients associated to the individual socio-economic characteristics.[5]

Following estimation of these parameters, the marginal rate of substitution (MRS) between any pair of attributes a and b can be calculated from the following equation:

$$MRS = - (\beta_{\text{attribute } a} / \beta_{\text{attribute } b}) \qquad \ldots (4.4)$$

Economic valuation of different scenarios can be estimated from a WTP measure derived from the following equation:

$$\text{Value} = - 1/\beta_m (V_0 - V_1) \qquad \ldots (4.5)$$

where the value represents an economic surplus derived from a monetary coefficient (β_m) and $V_0 - V_1$ represents the change in utility associated with changing from one state to another (Colombo et al. 2006), which in this study permits an understanding of the value of certifying land as a group of farmers compared to as an individual farmer.

Estimates from a multinomial logit model are presented in Table 4.4. Model I provides an attributes-only specification whilst Model II includes socio-economic characteristics.[6] All attributes are significant at the 1 per cent level and their signs are theoretically consistent. For example, a 1 per cent increase in crop price is associated with a positive value (0.111) whereas an additional labour day has a negative value (–0.041). The coefficient for committing an additional acre to organic farming is negative (–0.052), which is consistent with current farmer behaviour[7] and is associated with a higher level of negative utility from additional labour effort. The coefficient for compost price is also negative (–0.001). Land certification attributes are associated with positive utility and values are consistent with farmers preferring lower to higher certification costs and farmers deriving greater utility

---

[5]As respondents' characteristics do not vary across alternatives, singularities arise in model estimation unless the socio-economic characteristics are introduced as interactions with the attributes or alternative specific constants (Colombo et al. 2006).

[6]Though the socio-economic variables do not improve the attributes-only model to any great extent they are shown to be useful in a later specification (reported below).

[7]Percentage land commitments to organic farming in the choice cards (e.g. 25%, 50%, 75% and 100%) are estimated in acres based on respondent estimates of total cultivated land and existing area committed to organic farming captured in the questionnaire.

Table 4.4: Choice model results

		Model I[@]	Model II	Model III[#]	
				Class 1	Class 2
Attributes	Organic crop price increase (% increase)	0.111*	0.111*	0.096*	0.146*
	Land committed to organic farming (acres)	–0.052*	–0.052*	–0.481*	0.017*
	Labour effort for organic farming (days)	–0.041*	–0.041*	–0.001	–0.079*
	Compost price (Rupees per trolley)	–0.001*	–0.001*	–0.000	–0.001*
	Group land certification at Rs 1000 per acre	3.470*	3.446*	8.148*	3.438*
	Group land certification at Rs 3000 per acre	1.822*	1.796*	7.872*	0.912*
	Individual land certification at Rs 3000 per acre	1.537*	1.510*	7.458*	0.476*
	Constants[$]				
25% organic land	C1		–0.813*	–30.316	–0.364***
	Organic farm experience		0.190***	–0.790**	0.460*
	Head illiterate		0.034	–0.295	0.217***
	Income poor household		–0.042	–0.134	–0.183
	Head over 50 years		–0.055	0.720**	–0.295**
	Located in upper watershed		–0.007	0.562**	–0.267***
	Field sanitation		0.032	–0.250	0.186
	Poor house condition (kaccha)		–0.019	–0.639**	0.171
50% organic land	C2		–0.629**	–27.739	–0.530**
	Organic farm experience		0.120	–1.122**	0.549*
	Head illiterate		0.098	0.018	0.271**
	Income poor household		–0.137	–0.030	–0.278***
	Head over 50 years		0.074	0.778*	–0.013
	Located in upper watershed		0.057	0.776*	–0.346**
	Field sanitation		–0.028	–0.217	0.189
	Poor house condition (kaccha)		0.066	–0.222	0.174
75% organic land	C3		–0.620**	6.548*	–0.918*
	Organic farm experience		0.168	–0.340	0.243
	Head illiterate		0.170**	–0.375**	0.528*
	Income poor household		–0.129	–0.001	–0.060
	Head over 50 years		–0.076	0.247	–0.058

(contd...)

Table 4.4 (contd...)

	Model I[@]	Model II	Model III[#]	
			Class 1	Class 2
Located in upper watershed		0.167	0.170	–0.247***
Field sanitation		0.054	0.308	0.067
Poor house condition (kaccha)		0.054	–0.414***	0.241
Model summary				
Observations	5,116	5,116	5,116	
Log likelihood at convergence	–4530.68	–4511.99	–2509.09	
Log likelihood at constants	–6293.59	–6293.59	–6293.59	
Pseudo R^2	0.279	0.280	0.599	

Notes: * significant at the 1 per cent level; ** significant at the 5 per cent level; *** significant at the 10 per cent level.
[@] Independence from Irrelevant Alternatives (IIA) test, Chi-^{2} (7) = 15.15, p<0.05, choice 1 excluded; [#] Estimated probabilities of membership: Class 1 = 0.34*, Class 2 = 0.66*;
[$] Models do not report 100 per cent organic land as a choice constant to avoid dummy variable trap; organic field experience is defined by >10 per cent land cultivated organically; located in upper watershed dummy (UPK =1; LOWK and BMC = 0); all other dummy variables are as described in Table 4.3.
Source: Authors.

from working as a group rather than as an individual.[8] Model I has a significant model fit[9] supported by a pseudo R^2 value of 0.279.

In Model II, socio-economic characteristics are also estimated with a very small improvement in the model specification and little change in attribute coefficients. Only three of the four land commitment choices are reported to avoid the dummy variable trap. Constants are negative for each choice constant which is associated with a status quo bias. This is not uncommon in choice experiments and is associated with respondents' disutility in moving away from a current situation (Adamowicz et al. 1998). At least three reasons may be suggested for this outcome: (i) the choice task is too complex; (ii) there is uncertainty about the trade-off that a farmer would be willing to make; and (iii) farmers do not trust the government to implement the changes. It is uncertain which single or combination of factors has resulted in this outcome, though given that the respondents are small-scale

[8]This supports farmer comments from the scoping study and an understandable risk aversion to embarking on a significant and costly land conversion process alone.
[9]Likelihood ratio test = 3525.82, p<0.001

poor farmers, who are dependent on farming for food and income security, it is understandable that they may be cautious of committing to change without more information and consultation.

The slim pickings that are provided by the inclusion of socio-economic characteristics in Model II only suggest two indicators of organic farming adoption. At the 25 per cent level, only farmers with existing organic farming experience are likely to commit to organic farming and, at the 75 per cent level, only non-poor farmers are willing to commit such a large proportion of land to organic production. While these results may be expected, there are no other significant signals from the other socio-economic dummy variables. As the attributes-only model violates the Independence from Irrelevant Alternatives (IIA) test, it is appropriate to relax the IID assumption and consider more complex models that estimate preference heterogeneity (Boxall and Adamowicz 2002).

Latent class specification

A latent class specification introduces the additional complexity of preference heterogeneity along with the problem of assuming that the variance in the error term (e_i) is not constant but depends upon individual observations (heteroscedasticity). Commonly, a random parameters model is used to estimate heteroscedasticity in a continuous joint distribution (Louviere et al. 2000). Alternatively, a latent class model assumes a discrete number of support points (say, S) are sufficient to describe the joint density function. Latent classes correspond to underlying market segments or groups, each of which are characterized by unique tastes β_s, $s = 1,\ldots, S$. Swait (1994) notes that membership of class s can be characterized by variance differences across preferences, β_s, and scale, λ_s. If the indirect utility function for members of class s is:

$$U_{iq/s} = \lambda_s \alpha_{i/s} + \lambda_s \beta_s X_{iq} + \varepsilon_{iq/s} \qquad \ldots(4.6)$$

and the $\varepsilon_{iq/s}$ are conditionally IID extreme value type I within class, the choice probability for members of class s is estimated by:

$$P_{iq/s} = \exp(\lambda_s \beta_s X_{iq}) \bigg/ \sum_{j \in C_q} \exp(\lambda_s \beta_s X_{jq}) \qquad \ldots(4.7)$$

Swait (1994) provides details of a classification mechanism to predict an individual's membership in a class. The unconditional probability of choosing alternative i is given by:

$$P_{iq} = \sum_{s=1}^{S} P_{iq/s} W_{qs} \qquad ...(4.8)$$

where W_{qs} is the probability of being in class s.

An advantage of using a latent class specification, in this context, is that an understanding of farmer preference heterogeneity can be estimated without arbitrary decomposition by income, land holding, or zone (as in the descriptive analysis), thus potentially revealing classes (or groups) of farmers hidden within the aggregate model and identified by an associated probability of class membership.

Model III in Table 4.4 presents results from a latent class specification. Of particular note is the large improvement in the model fit from a pseudo R^2 of 0.280 in Model II to 0.599 in Model III. Two classes, which are significant at the 1 per cent level, are revealed with Class 1 having an estimated 34 per cent probability of farmer membership and Class 2 having an estimated 66 per cent probability of farmer membership. Attribute coefficient estimates also differ markedly from Model II attributes and the socio-economic characteristics now provide significant estimates of farmer choice preferences.

In Class 1, the attributes for land commitment to organic farming and labour effort are not significant. The utility from a 1 per cent increase in crop price is lower (0.096) in Class 1 but higher in Class 2 (0.146) compared to the aggregate model.[10] This indicates a higher price preference for Class 2 farmers. Of particular note, is the change in the land commitment for Class 2 farmers, which is positive though small. While this may be unexpected, it is balanced by a high disutility of labour effort (-0.079) which is almost double the aggregate estimate (-0.041). This suggests that while Class 2 farmers associate positive utility with converting land to organic production, this process is only likely to be achieved through a higher price incentive and a low farmer labour investment. Land certification attributes also reveal significant changes from aggregate model estimates. While Class 1 farmers record high positive utility estimates for certifying land across the three scenarios, price per acre has little impact on utility levels. In contrast, Class 2 farmers record an almost identical utility estimate to the aggregate score for group certification at Rs 1000 per acre (3.438) but reveal a far larger utility improvement from this option in comparison to the higher

---

[10]From here, aggregate model refers to Model II.

Table 4.5:   Marginal values and implicit prices

	Model II	Model III	
		Class 1	Class 2
Marginal rate of substitution			
One acre of organic farm to a 1% increase in crop price	2.13	0.20	n/a
One labour day to a 1% increase in crop price	2.71	n/a	1.85
Certifying land as a group rather than as an individual (cost same)	1.93	1.04	3.77
Implicit prices*			
1% increase in crop price	R111	n/a	R146
Land certification as a group at Rs 3000 per acre	R1,796	R7,872	R912
Land certification as an individual at Rs 3000 per acre	R1,510	R7,458	R476
Net marginal benefit of certifying land as a group per acre	R286	R414	R436

Notes: *Implicit prices estimated from compost price estimate.
Source: Authors.

cost certification options. These findings are presented in terms of implicit prices and marginal rates of substitution in Table 4.5.

Marginal rates of substitution indicate that a 1 per cent increase in crop price will lead to 2.13 acres committed to organic production for Model II farmers compared to only 0.20 acres for Class 1 farmers, which is consistent with their high disutility of committing land to organic production.[11] In terms of labour commitment to a 1 per cent crop price rise, Model II predicts 2.71 additional labour days in contrast to 1.85 days for the labour-sensitive Class 2 farmers. Farmer sensitivity to working collectively or individually reveals Class 1 farmers as least willing to work together (1.04), Model II farmers show greater willingness to certify land in groups (1.93), and Class 2 farmers with the strongest desire to work together at an equal cost of Rs 3000 per acre (3.77). Implicit prices suggest a 1 per cent increase in crop price is valued at Rs 111 by Model II farmers compared to Rs 146 by Class 2 farmers. Finally, implicit net benefit is estimated for farmer preferences to working collectively to certify land at Rs 286 per acre for Group II, Rs 414 per acre for Class 1 and Rs 436 per acre for Class 2.

[11]Given two positive estimates for price and land for Class 2 farmers, the sign of the MRS is negative.

The latent class specification also provides insights into socio-economic characteristics of Class 1 and Class 2 farmers across land commitment scenarios. The characteristics of Class 1 and Class 2 farmers committing to 25 per cent organic farming are almost asymmetrical. For example, a Class 1 farmer would be characterized by no organic farming experience, be over 50 years old, live in the upper watershed and in a non-poor home. Conversely, a Class 2 farmer would have organic farming experience, be younger, live in the lower watershed, and be illiterate. Characteristics for committing 50 per cent land indicate that Class 1 farmers again have no experience of organic farming, are over 50, and live in the upper watershed compared to Class 2 farmers, who do have organic experience, are illiterate, are non-income poor, and live in the lower watershed. Finally, the profile for 75 per cent land commitment suggests Class 1 farmers are literate and live in non-poor homes.[12] In contrast, Class 2 farmers are illiterate and live in the lower watershed.

Conclusion

Improved soil and water management is a significant policy challenge in tropical countries in order to conserve rapidly degrading environmental resources and reduce entrenched rural poverty. These two objectives are often at odds due to a failure to understand the competing interests and impacts of inter-dependent and complex biophysical and social system interactions. Identifying realistic approaches that are socially-acceptable and environmentally-beneficial requires objective information on public preferences for policy scenarios. This chapter has explored such preferences in relation to the feasibility of small-scale farmers' adoption of organic land management in an ecologically-sensitive watershed in India.

A choice experiment method has been found to be a suitable approach to estimate farmer preferences to inform policy design. Of particular note, in the rural context of this study, is how the method has been able to combine earlier qualitative inquiry with primary stakeholders (such as farmers or land managers) and secondary stakeholders (for example, government, wetland users) to design experimental scenarios that can be pictorially presented to some illiterate stakeholders, who may otherwise be excluded from decision

[12]The sign of the constant is positive and significant suggesting there is no status quo bias.

making. This has important participatory benefits which are consistent with the democratic governance in India, and are more widely thought to be critical to achieving sustainable development (MA 2005). Though incentives are considered central to shifting farmer land management behaviour, a status quo bias suggests the choice task proved too difficult for some respondents or there was limited trust in effective implementation of the choice scenarios.

It is likely that continuing farmer use of agro-chemicals in the recorded quantities will further degrade the Bhoj wetland ecosystem and have implications for municipal water supply and other wetland users (Mishra 2006). However, based on this analysis, there is evidence that with the right incentives and institutional support, farmer land management decision making can be changed. A farmer group (Class 2) is identified as a promising target for adopting organic land management based on a positive willingness to commit up to 50 per cent of land to organic farming and a strong preference to work collectively in the institutional process of land certification. However, Class 2 farmers prefer a higher crop price incentive compared to the watershed sample of farmers and are less likely to participate if organic farming is labour-intensive. This information is useful in determining the viability of wider organic farming extension activities. Equally, results illustrate the profile of Class 2 farmers to be characterized as located in the lower watershed zones (LOWK, BMC), to have experience of organic farming, to be under 50 years of age, and to be illiterate. A second group of farmers (Class 1) appear a less promising target for adopting organic farming. However, at the higher level of 75 per cent commitment, literate farmers in non-poor homes appear likely to adopt organic farming at a lower price incentive, though they appear unlikely to certify land collectively. Importantly, this choice constant does not register a status quo bias, which may be associated with literacy and a higher welfare proxy. These results provide guidance to the 'who, where, and how' organic farming may be adopted in the Kolans watershed.

Given that shifting to organic farm management is likely to be more profitable than current land use, off-site wetland benefits will be contingent on effective implementation of organic land certification. This suggests that if farmers can access premium organic markets, land certification may provide a self-enforcing institutional arrangement for land management change. Such an arrangement will require transitional financing and capacity-building for farmers shifting to organic production and thus reduce long-term transaction costs and monitoring of off-site water quality impacts from agricultural run-off. Policy makers need to consider further steps to effectively implement such an integrated intervention, including securing transitional financing, organic market and price analysis, identifying compost supply, institutional support in farmer land certification, and building farmer capacity.

REFERENCES

Adamowicz, W., P. Boxall, M. Williams, and J. Louviere (1998), 'Stated Preference Approaches for Measuring Passive Use Values: Choice Experiments and Contingent Valuation', *American Journal of Agricultural Economics*, Vol. 80, pp. 64–75.

Bassi, L. (2002), *Valuation of Land use and Land Management Impacts on Water Resources in Lajeado Sao Jose micro-watersheds, Chapeco, Santa Caterina State, Brazil,* Land-Water Linkages in Rural Watersheds Case Study Series, FAO, Rome.

Boxall, P.C. and W.L. Adamowicz (2002), Understanding preference heterogeneity in random utility models: A latent class approach, *Environmental and Resource Economics*, Vol. 23, pp. 421–46.

Borgoyary, M. (2005), *Scoping report: Negotiating Socio-economic Opportunities in Payments for Environmental Services schemes, the case of the Bhoj wetland,* Unpublished project report (FRP R8174), Winrock International India.

Colombo, S., J. Calatrava-Requena, and N. Hanley (2006), Analysing the social benefits of soil conservation measures using stated preference methods, *Ecological Economics*, Vol. 58, No. 4, pp. 850–61.

Deaton, D. and J. Dreze (2002), Poverty and Inequality in India—A Re-examination, *Economic and Political Weekly,* 7 September 2002, pp. 3729–48.

Environment Agency (2002), *Agriculture and Natural Resources: Benefits, Costs, and Potential Solutions*, Environment Agency, Bristol.

Ghosh, N. (2004), Reducing dependency on chemical fertilizers and its financial implications for farmers in India, *Ecological Economics,* Vol. 49, pp. 149–62.

GoI (2005), *Some Aspects of Farming: Situation Assessment Survey of Farmers,* NSS 59th Round, National Sample Survey Organization, Report No. 496 (59/33/3), Government of India, July.

Harris, P.J.C., A.W. Browne, H.R. Barrett, and K. Cadoret (2001), *Facilitating the Inclusion of Resource-Poor in Organic Production and Trade: Opportunties and Constraints Posed by Certification,* Unpublished DFID report, Henry Doubleday Research Association and Coventry University, UK.

Hausman, J.A. and D. McFadden (1984), Specification tests for the multinomial logit model, *Econometrica*, Vol. 52, pp. 1219–40.

Hope, R.A., M. Borgoyary, and C. Agarwal (2005), *Designing a Choice Experiment to Evaluate Adoption of Organic Farming for Improved Catchment Environmental Services and Poverty Reduction*, Unpublished project report, University of Newcastle, UK.

IUCN (n.d.), *Regional Wetlands and Water Resources Programme (RWWP) in Asia*, World Conservation Union, Bangkok.

Lancaster, K. (1966), A new approach to consumer theory, *Journal of Political Economy*, Vol. 74, pp. 132–57.

Louviere, J.J., D.A. Hensher, and J.D. Swait (2000), *Stated Choice Methods: Analysis and Application*, Cambridge University Press, Cambridge.

MA (2005), *Ecosystems and Human Well-Being: Wetlands and Water. Synthesis*, Millennium Ecosystem Assessment, World Resources Institute, Washington, D.C.

McFadden, D. (1973), Conditional logit analysis of qualitative choice behaviour, in P. Zarembka (ed.), *Frontiers in Econometrics*, Academic Press, New York, pp. 105–42.

Mishra. S.M. (2006), *Impact of the Rural Catchment on the Bhoj Wetland and Options for Mitigation*, Interim report, Lake Conservation Authority of Madhya Pradesh, Bhopal.

Pagiola, S., P. Agostini, J. Gobbi, C. de Haan, M. Ibrahim, E. Murgueitio, E. Ramirez, M. Rosales, and J.P. Ruiz (2004), Paying for Biodiversity Conservation Services in Agricultural Landscapes, *Environment Development Paper*, Vol. 96, World Bank, Washington, D.C.

Stolze, M, A. Priorr, A. Haring, and S. Dabbert (2000), *The Environmental Impacts of Organic Farming in Europe, Organic Farming in Europe, Economics and Policy*, Vol. 6, Stuttgart, University of Stuttgart-Hohenheim, Stuttgart, Germany.

Swait, J. (1994), A structural equation model of latent segmentation and product choice for cross-sectional revealed preference choice data, *Journal of Retailing and Consumer Services*, Vol. 1, No. 2, pp. 77–89.

van Noordwijk, M., F. Chandler, and T.P. Tomich (2004), *An Introduction to the Conceptual Basis of RUPES: Rewarding Upland Poor for the Environmental Services they Provide*, ICRAF-Southeast Asia: Bogor, Indonesia.

Verma, M. (2001), *Economic Valuation of the Bhoj Wetland for Sustainable Use*, Unpublished project report for World Bank assistance to Government of India, Environmental Management Capacity Building, Indian Institute of Forest Management, Bhopal.

World Bank (2003), *India: Sustaining Reform, Reducing Poverty*, World Bank, Washington, D.C.

Wunder, S. (2005), *Payments for Environmental Services: Some nuts and bolts*, CIFOR Occasional Paper No. 42, Bogor, Indonesia.

5

Measuring Economic Value of Crop Pollination Service
An Empirical Application of Contingent Valuation in Kakamega, Western Kenya*

MUO J. KASINA, JOHN MBURU, AND KARIN HOLM-MUELLER

INTRODUCTION

Although pollination is a service of nature that has been grouped together with other services which provide regulatory functions in the ecosystem (de Groot et al. 2002), it has increasingly been recognized as a very significant contributor to the agricultural production of a broad range of crops, particularly fruits, vegetables, fibre crops, and nuts (Olmstead and Wooten 1987; Kevan and Phillips 2001). In agricultural systems, pollination is viewed as an important input for crop production, comparable to fertilizers or pesticides, as it is one of the determining factors for successful crop reproduction. Yet, even with this importance, animal pollinators, which are key agents of pollination for some particular crops, are reportedly declining rapidly particularly due to factors associated with human influence or neglect Convention on Biological Diversity (CBD 2001). An indirect factor for this decline is also the fact that the economic value of pollination is least understood or appreciated by policy makers and other stakeholders in developing economies.

There have been several attempts in the past to quantify the contribution of pollination to agriculture. The main underlying problem, however, has been the nature of its provision. Pollination has public good characteristics and in many countries, particularly

*This study was conducted within BIOTA East Africa project. The authors greatly appreciate the financial support of the Germany Federal Ministry of Education and Research (BMBF).

developing ones, it is provided 'free' by a healthy ecosystem with no human (farmer) intervention. This implies that the good is freely accessible, collectively consumed, and indivisible among the farmers. However, in some developed countries, bees have been reared, managed, and traded for pollination purposes and as such the pollination service has a market value. Studies to measure the economic value of pollination have been conducted in such countries where pollination markets exist for example, in USA (Southwick and Southwick 1992), UK (Carreck and Williams, 1998), and Australia (Gordon and Davis, 2003). These studies mainly use revealed preference (RP) methods to value pollination (see Freeman 2003). Since RP methods involve indirect valuation through observation of market transactions, it is possible to infer the economic value from the prevailing market price of pollination service or from the additional value of crop commodities realized through enhanced pollination. Although cost-based methods such as preventive expenditure, damage costs avoided, and replacement cost can be used to value pollination service, there has been little application even in developed countries (Hein and Mburu 2006). Similarly, stated preference (SP) methods, such as contingent valuation Contingent Valuation Method (CVM) have not been widely applied mainly due to the limitation that respondents may not have a sound knowledge of the quantitative contribution that pollination makes to agricultural production in their fields. They have, however, been used to value many kinds of ecosystem goods and services where pollination is included as a part of the ecosystem but not on its own (for example, Costanza et al. 1997).

As a modest attempt to value pollination in a developing country and apply SP methods, this study measured the value of pollination service from the perspective of local farmers by using CVM in Kakamega district, Western Kenya. The method was applied after sensitizing farmers of the importance of pollination and making sure that they understood its contribution to crop production in their farms. This method was also deemed suitable due to the fact that pollination service in the country is not traded in the market although it is quite important considering the fact that different kinds of crops grown in the research area require animal pollination.

The findings of this study are expected to open up a public debate among stakeholders (for example, farmers, policy makers, and civil society) on the economic role of pollination in the research area and other parts in the country. The research area is of particular importance due to its proximity to the Kakamega forest, one of the main sources of pollination service for crops grown in the surrounding farmlands (BIOTA 2004). It is also unique in that the area is a high potential agricultural zone and the inhabitants are dependent on staple crops, vegetables, and fruit crops that rely on biotic pollination for their well-being.

The rest of this chapter is organized as follows: in the next section, we provide information about the study area and sampling procedures. In the third section we describe the methodological aspects, while in the fourth section we discuss the results. Finally, in the fifth section, we draw conclusions and provide a few policy implications.

STUDY AREA AND SAMPLING DESIGN

Study area

Kakamega is situated in the western region of Kenya (Map 5.1). The area has rich agricultural soils, plenty of rain (more than 1500 mm per annum), which is well distributed throughout the year, and an average temperature of 22°C. These attributes make it suitable for farming, and it is classified as one of the high potential areas for agricultural production in Kenya (Jaetzold and Schmidt 1982). The populace consists of small land units that are reported to be as low as 0.2 ha (Greiner 1991), and which continue to be subdivided to smaller units. The farmlands are highly populated with a range of 433 to 713 inhabitants per sq. km having been recorded (Greiner 1991; BIOTA 2004), making it one of the most populated rural dwellings in the world. Most households are small-scale farmers of crops and livestock. They rely mainly on staple crops, vegetables, and fruit crops (such as legumes, solanaceous crops, cucurbits, avocado, and passion fruits) for their daily dietary requirements. Most of these crops require biotic pollination and are highly affected by declining pollination service in the research area.

One of the unique features of the research area is the presence of the Kakamega forest, which is exceptional in that it has a mixture of Guineo-Congolian and Afro-montane fauna and flora species. It is also the only remaining patch in Kenya of the once Guineo-Congolian rainforest that used to span from West Africa through Uganda to Kenya. It is the most species-rich forest in Kenya and has large numbers of rare animal and plant species, some of which are endemic (KIFCON 1994). The forest is one of the main sources of pollination

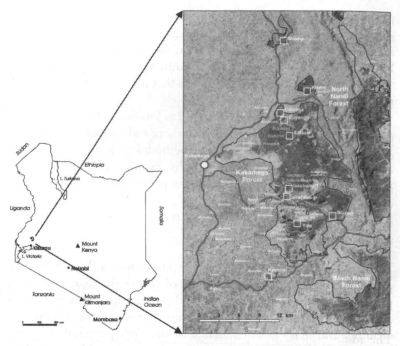

Map 5.1:   Map of Kenya showing location of Kakamega forest

Note: Landsat ETM + (7) satellite image (5 February 2001, spectral bands 5/4/3, contrast enhanced) of Kakamega Forest, its peripheral fragments and the adjacent Nandi forests. White frames mark 10 BIOTA study sites
Source: USGS; access and preparation: BIOTA-E02, G. Schaab, Karlsruhe, Germany.

service to the crops grown in the farmland but due to continued disturbances and intensive farming, this pollination service provision is being threatened (BIOTA 2004). If measures are not taken to conserve pollinators, there is a likelihood of high reduction in commodities produced from pollinator-dependent crops. On the other hand, measures needed to support improvement of population size of the pollinators would demand changes in the farmland landscape. Such measures would also impact the Kakamega forest, in halting its degradation. Households neighbouring the forest usually create a buffer zone between the forest and themselves. In essence, they have gains (for example, pollination), as well as conflicts (for example, wildlife–human conflicts such as crop damage), from the forest. Therefore, any intervention for pollinator conservation in the farmland would impact positively on the adjacent forest and can ensure continued provision of other services accruing from it.

Sampling design

The study was conducted within the farmland surrounding the Kakamega forest. It included households living up to a distance of 10 km from the edge of the forest. All households living within this perimeter were documented just before the beginning of the enumeration process. A total of 19,972 households in 210 villages were listed as the target population. For reasons of cost effectiveness and better questionnaire administration, multi-stage random sampling was done on two levels to get a sample of 352 households to be interviewed. Thus, 22 villages were randomly selected (first level) and for each village 16 households randomly sampled (second level). This size of the sample was deemed large enough to take care of the protest bids. A formal questionnaire was developed to generate information about the socio-demographic characteristics of the households, respondents' perception of pollination as a crop production input, and their WTP for enhanced pollination service. The administration of the questionnaire was done through face-to-face interviews in January and February 2006. Four university graduate enumerators who understood the local language and had enumeration experience from previous studies were used to conduct the interviews. They were first trained intensively on the enumeration process and study objectives. A pre-test survey was done to assess how well the survey worked as a whole, to improve the language used in the narrative, and to help to fine-tune the questions. Usually, a village elder was approached to introduce the enumerators to the respondents in each village. This made the respondents feel comfortable with the enumerators. Generally, the respondents were hospitable, receptive, and interested in the survey, which was shown by the high number of positive responses compared to the low number of non-responses received.

METHODS

Contingent valuation

Two main distinctions can be drawn between the methods used to value ecosystem goods/services, depending on the source of the data required for the assessment (Mitchell and Carson 1989; Pearce and Turner 1990; Freeman 2003). The data may be taken indirectly from the regular marketplaces (for marketed goods) or directly from the consumers using prepared set of questions (for non-marketed goods). By observing the market behaviour, information is indirectly revealed by the consumers about their preference for the good (hence the RP methods). On the other hand, asking the consumers how much they would be willing to pay for the good implies that they state their preferences directly (hence the SP methods). Thus, SP methods rely on information given by the consumer about the good in question. They are, therefore, the most appropriate for valuing an ecological service such as pollination, which in the Kenyan context is non-marketed.

CVM is one of the SP methods used in valuation studies, especially in the case of non-marketed ecosystem services (see FAO 2000; Perman et al. 2003). It uses a direct approach in valuing the good by asking the consumer to state his/her WTP to have it, or willingness to accept compensation (WTA) for the loss of the same. This method, therefore, relies on a hypothetically constructed market where individuals are allowed to make transactions. The contingent market provided includes the good itself, the institutional context on how it will be provided, and how it will be financed (Carson et al. 2001; OECD 2002). The main aim is to achieve values close to those that would be revealed in an actual market if it existed. The contingent market is, therefore, the central tool in SP valuations.

The CVM has been used in many countries successfully to measure the value of different ecosystem services and biodiversity (for example, Bandara and Tisdell 2004; Ellingson and Seidl 2007; Tisdell et al. 2007). It has not been used in agriculture unlike other SP methods such as choice modelling due to reasons mentioned earlier.

Valuation context and empirical procedures

Households in Kakamega farmland require several inputs of crop production in every cropping season. Some of these inputs are usually purchased in the local market, and their value can easily be inferred from the market transactions (indirect measurements). Other inputs (such as pollination services) are not available in the market and are, therefore, not purchased. However, households benefit from these services because they are provided 'free' by nature. In such a case, pollination is a positive externality of forest conservation and landscape management activities in their current production system. The service is used in production but farmers don't factor it in their cost appropriation. It is a purely public good in the current situation at Kakamega, but the provision by nature has been declining mainly due to reduction in bee population sizes (BIOTA 2004; Gikungu 2006). Therefore, a direct approach such as CV is the appropriate tool to measure the value of pollination to these households (Mitchell and Carson 1989, Perman et al. 2003). The choice of either WTP or WTA literally depends on the ownership rights of an economic agent (households). We assume here that farmers in Kakamega have the right to the current level of provision of pollination service, but they have to pay if they want improved services (more of the good). Therefore, they were asked to state their maximum WTP to have the new or enhanced improved pollination service.

In economic terms, WTP can be described as the amount of money that can be deducted from the respondent's income when increasing the provision of the public good, keeping his/her utility constant. We assume that the respondents' utility function is represented by:

$$U = (x, q_i, l),$$

where x is the vector of quantities of market goods, q is the 'quantity' of ecosystem good (herein pollination) with level i, and I is income. The utility maximization problem can be expressed as $\underset{x}{Max}(x, q, I)$ such that $px = I$ where p is the vector of prices. The solution to this problem leads to a set of demand function, $x^* = x(p, q, I)$. The indirect utility function is then defined as $V(p, q, I) = U(x^*, q, I)$ with $V_q > 0$ and $V_I > 0$ (see Binger and Hoffmann 1988 for detailed description and derivation of the indirect utility function).

This study values pollination change, that is, improved pollination service (higher quantity of the pollination good). The decrease in income that keeps an individual at the same utility level before the change in pollination 'quantity' is the measure of welfare impact. This indifference point can be measured as:

$$v\ (p, q_0, I) = v\ (p, q_1, I - y) \qquad \qquad ...(5.1)$$

where 0 and 1 are the original (0) and improved (1) level of pollination, q, respectively, and y is the maximum WTP value (the value that was elicited from the respondents by this study). The utility of the respondent was also hypothesized to depend on the individuals' characteristics that influenced the trade-off a respondent was prepared to make between income and pollination quantity. Therefore, a vector, Z, representing the characteristics of an individual can be added to equation 5.1 (see also FAO 2000):

$$v\,(p,\,q_0,\,I;\,Z) = v\,(p,\,q_1,\,I - y;\,Z)$$

In this study, the WTP values provided by the respondents were averaged to produce an estimate of the sample WTP, because the payment question was open-ended. This can be expressed as:

Sample $\qquad\qquad WTP = \dfrac{1}{n}\sum_{i=1}^{n} y_i$

where n is the sample size and y is the reported WTP amount by a respondent, i.

The median value is sometimes the most relevant in the presentation of the economic gain from the public good. Policy makers are keen on the median value, as it represents the amount many people are likely to pay, compared to the mean value, which is sometimes prone to influence by extreme values (and/or outliers) given by some respondents. However, median values are most often presented in CV exercises that use referendum-format questions compared to open-ended formats (Mitchell and Carson 1989).

The values of WTP responses given by the respondents in this study were checked to see whether they relate in a predictable way and to find out which variables influenced the magnitude of the WTP values. This was done by regressing the reported WTP values (regressands) against the socio-economic attributes (regressors) of the respondents by using the Tobit model (Tobin 1958).

The Tobit model was devised by Tobin (1958) and assumes that the dependent variable has a number of its values clustered at a limiting value, usually zero. The model uses all observations—those at the limit as well as those above it, to estimate a regression line. It is usually the most preferred model over others, which estimate the regression line only with the observations above the limit. The stochastic model underlying the Tobit model can be expressed by the following relationship:

$$y_i = \begin{cases} y_i^* & \text{if} & y_i^* > 0 \\ 0 & \text{if} & y_i^* \leq 0 \end{cases}$$

where y_i is the observed dependent variable, and y_i^* is the unobserved latent variable, which is expressed as:

$$y_i^* = \beta X_i + u_i \quad \text{where } u_i \sim N(0, \sigma^2)$$

where X_i is the vector of the independent variables, β is the vector of unknown coefficients, N is the number of observations, and u_1 is an independently distributed error term assumed to be normal with zero mean and constant variance σ^2.

This model assumes that there is an underlying stochastic index equal to y_i^* which is observed only when it is positive, and hence qualifies as an unobserved, latent variable. This latent unobservable variable y_i^* is linearly dependent on X_i through β. The β parameter, therefore, determines the relationship between the independent variable (or vector) X_i and latent variable y_i^*. The normally distributed error term u_i captures the random influences on this relationship. The observable variable, y_i, is equal to the latent variable, y_i^*, whenever y_i^* is above zero but it is equal to zero when y_i^* is equal to, or less than, zero.

Estimation of the relationship parameter β by use of the ordinary least squares (OLS) method is usually inconsistent compared to the likelihood estimation obtained using the censored Tobit model (see McDonald and Moffitt 1980). In this study, the regressand, WTP (y_i), is known (observed) for some respondents (who gave a non-zero value), but this is not so for other respondents (who gave zero values). The assumption is that the y_i of those who gave the zero value was not observed, hence we use y_i^*. However, regressors (X_i) for the two groups are available, thus making the Tobit model the most suitable to carry out the regression analysis.

Socio-economic factors were recorded during the enumeration process based on the previous experience of other CV studies elsewhere in the world, and the mode of payment vehicles used by this study (see section titled 'Payment Vehicle for WTP' for the payment vehicles used in the study). For example, income and education have been reported by many studies to influence the magnitude of the WTP value (for example, Boyle and Bishop 1987;

Heinen 1993). It was expected that educated individuals would be willing to pay more, as they may appreciate the role of pollination service more than the less educated. Large households were expected to provide more days of labour, as they may have spare labour compared to small-sized households. Older respondents and females were expected to provide meals due to lack of energy and time, respectively. A negative correlation between age and education was also expected, since most CV studies done in developing countries show that older people are less educated than younger ones (see FAO 2000). It was expected in this study that young people would pay more for the pollination service than the older ones. Religion was also expected to impact the choice of payment vehicle as most Catholics in the farmland usually attend weekly prayer meetings and other associations which could constrain their payment. In the research area, women are the main household members involved in farming activities and were expected to have a better understanding of the impact of the enhanced pollination service to the crop yields especially after sensitization exercises. It was also expected that households nearest to the forest would pay less because of their proximity to the forest, which is the main habitat for the bee pollinators of crops.

The contingent market and elicitation method

Before the introduction of the WTP scenario and questions, background information had to be sought to find out the level of understanding of the respondents on the good being valued (bee pollination service) and provide the missing information about the good to the respondents. Farmers that were not aware of, or did not know, the importance of pollination were, therefore, first sensitized about the service before the hypothetical market and other questions were presented to them.

The scenario that was used to provide the pollination service to the consumers (respondents) was designed such that the hypothetical market presented would be as close as possible to the real market (see Carson et al. 2001) and would appeal to the respondents to make transactions. This is important, as it renders the WTP results reliable and useful for use by policy makers. Considering the target group, this study used a nature conservation project where the respondents were asked to pay for conservation of bees in order to guarantee better pollination of their crops. The hypothetical market consisted of a planned pollinator garden where such activities such as construction of nesting, feeding, resting, and hiding sites of different bee species were to be carried out. Also included was the construction and maintenance of corridors for bees in the farmland. The respondents were to participate in such duties. This scenario was pre-tested and qualified as the best model to value pollination service as an input of crop production in the region using the CV technique.

In order to guarantee the validity of the valuation exercise, the scenario was described and presented as clearly as possible to each respondent using the local language. This involved not only careful description and explanation, but also graphical visualization, for example, the use of pictures, graphs, and illustrations. Open-ended questions were used to elicit the maximum amount each respondent was willing to pay for the pollination service. This format has been used successfully by other studies elsewhere to value public goods (for example, Tanguay et al. 1995; Allport and Epperson 2003). The stipulation of this contingent market was done cautiously so as not to overwhelm the respondent with a lot of information and cause confusion (see OECD 2002).

After they had listened to the scenario, but prior to the introduction of the payment question, respondents were asked whether they were willing to pay for the pollination service in the hypothetical market. This enabled them to register protest bids or their true zero WTP value by refusal to pay for the pollination service without feeling uneasy or intimidated in doing so. Those respondents who agreed to pay were then asked to state the preferred payment vehicle out of the two alternatives presented to them. This was followed by elicitation of the maximum amount they were willing to pay; those who did not accept were asked to give reasons for their action.

Payment vehicle for the WTP

Kahneman and Knetsch (1992) indicated that the most important criterion for the success of the method of payment is that it should be believable. For this study, two vehicles of payment were chosen and presented as clearly as possible to the respondents. The respondents were asked either to provide own labour up to midday (half a day) to the nature conservation project (pollinator garden) or lunch-time meals to the workers employed in the project. Both vehicles were expressed on a weekly base either in half-days of labour or number of lunch-time meals provided. The respondents were not allowed to pay in cash, because they are oversensitive to this mode of payment and would always complain about lack of money. The meal or labour payment vehicles were selected because they are common transactions in the research area and are not new to the respondents. Normally, casual labourers in the research area have an alternative to work for food or money. When a lunch-time meal is provided, the wage rate is US$ 0.67 (Kenya shillings, Ksh, 50.00) per day, but this amount doubles when there is no request for food. Thus, the respondents were reminded that the cost of one day of labour up to midday (one half day of labour) is equivalent to the cost of a single lunch-time meal and each amounts to US$ 0.67.

It was also emphasized that the only gain from the payment was improved pollination of their crops and the expected crop yield gain. Respondents were told explicitly that the payment was weekly and continuous (long-term) in order to restore and ensure long-term maintenance of pollinators and hence the pollination service. This helped to eliminate the temporal embedding problem (Carson et al. 2001), as the respondents knew the commitment required of them if they agreed to pay. Respondents were reminded of this before the WTP question. We addressed the bias that respondents could provide answers for all crop pollinators rather than just bees by the adequate survey design that we had developed and the information that we provided during the interview session in a simple and logical sequence (see Carson et al. 2001).

Data analysis

LIMDEP-NLOGIT 3.0 software was used to run the Tobit regression and perform the analysis of the descriptive statistics of the survey respondents. Means and frequencies of the different descriptive parameters are provided. Standard deviation (SD) from means was used to separate means and significance of the means was tested at different levels.

Results and Discussion

Socio-economic characteristics and awareness of pollination

The median and mean age of the sample respondents are almost equal (Table 5.1), implying that the distribution of age is not skewed but normally distributed. Many respondents were household heads/ spouses (85 per cent) and married (81 per cent). This is an added advantage to the reliability of the survey results since the respondents, as the main decision makers, would consider household constraints while stating their WTP. About 61 per cent of the respondents were female. In the research area, women are the main household members involved in farming activities and understand better the impact of an enhanced pollination service to the crop yields. Furthermore, 66 per cent of the respondents had more than six years of formal education. The respondents showed uneasiness when revealing their income but not when asked about their expenditures. Therefore, expenditure was used as a proxy to estimate income. The mean monthly household expenditure was US$ 110.07.

While most (99 per cent) respondents knew about bees, only about half of the sample (47 per cent) knew about the function of pollination in crop production prior to the sensitization (Table 5.1). After explaining the process of pollination and the role of bees in pollination and hence crop yield gain, respondents were asked again to state whether they understood what pollination was and its importance as an input of crop production. Most of the respondents (84 per cent) were affirmative, stating that pollination was very important to them. It is, however, important to note that even those who did not understand its role (mainly due to low education level) were willing to support its conservation after being told that the service exists to support crop production. Thus, the WTP of the respondents (98 per cent) for the pollination service may be due to, among other factors, their appreciation of its role in crop production and their knowledge of its mere existence. The importance of pollination in the region was given extra weight by the indication that most respondents were growers of crops that require animal pollination. Bee pollination, therefore, was well perceived as a contributing factor to crop yield gain and a worthwhile investment to enhance crop production.

Table 5.1: Demographic and socio-economic characteristics of the sample

Variable	Description	Mean	SED
GENDER	1 male; 0 female	0.38	0.489
MARITAL	1 married; 0 otherwise	0.81	0.392
AGE	Age in years	41.00 (median: 40.00)	15.765
EDU	Years of formal schooling	6.27	4.083
HH_S	Number of individuals in a household	5.88	2.497
HHD_S	1 Household head or spouse; otherwise	0.85	0.361
EXP	Mean monthly expenditure per household (US $)	110.07	88.372
CG	A crop grower (1 yes; 0 no)	0.99	0.076
LSK	A livestock keeper (1 yes; 0 no)	0.94	0.229
KOB	Having knowledge of bees (1 yes; 0 no)	0.99	0.093
KOP	Having knowledge of pollination (1 yes; 0 no)	0.47	0.499
PERC_POL	Perception of pollination(1 very important, 0 otherwise)	0.84	0.409
OFF_INC	Having off-farm income generating activities (1 yes; 0 no)	0.58	0.494
RELIGION	Religious affiliations (1 catholic, 0 otherwise)	0.32	0.462
DIST_FOR	1 household within 5 km range from forest edge; 0 more than 5 km	0.55	0.491
Willing to pay	Willing to pay for pollination service (1 yes; 0 no)	0.98	0.131
PV	Preferred payment vehicle (1labour; 0 meal)	0.65	0.476
Labour WTP	Willingness to pay value (weekly half-days of labour)	2.53	1.154
Meal WTP	Willingness to pay value (weekly number of lunch-time meals)	2.62	1.356
Sample WTP	Willingness to pay value (weekly half-days of labour or number of lunch-time meals)	2.52 (median: 2.00)	1.260

Notes: N = 352 for the general descriptives and 345 for the willingness to pay and payment vehicle values; US $ 1.00 = KSh. 75
Source: Authors.

Willingness to pay

Out of the 345 respondents, 98 per cent were willing to pay for the pollination service (Table 5.1). The high number of respondents who were willing to pay suggests that they felt that enhanced pollination service would contribute positively to their well-being. It also confirms that the market scenario presented was well understood and appealed to them. Among those who were willing to pay, 65 per cent chose labour as the preferred payment vehicle. Since the lunch-time meals cost the same as half-day labour (US$ 0.67), the choice of labour by a majority of the respondents did not necessarily imply that this payment vehicle was cheaper. Rather, this choice must have been influenced by other factors, some of which we try to capture with econometric model results (Table 5.2).

Table 5.2:   Reasons given by respondents for refusing to pay for enhanced pollination service

Reason	No. of respondents	Zero value
I keep honeybees	1	True zero
I have no time to participate	4	True zero
I fear more charges at later stages	1	True zero
The household head has to determine	4	Protest zero
I am old and have no energy	3	Protest zero

Source: Authors.

Respondents who didn't participate in the valuation exercise (who refused to pay for improved pollination) gave varying reasons, which enabled their classification as either true zero value of pollination or a protest zero (Table 5.2). Respondents who gave protest values were excluded from the estimation of the WTP function unlike those who gave a true zero value for improved pollination service. The protest was probably due to the unconvincing market scenario, which did not satisfy them.

Respondents were reminded of the temporal dimension of their payment and were requested to take into consideration their daily schedules and budgetary constraints when stating the WTP value. This criterion is important in order to reduce temporal bias (Carson et al. 2001; Freeman 2003). The mean household WTP per week was 2.53 half-days of labour and 2.56 lunch-time meals (Table 5.1).

After including the true zero values of the respondents, the average household WTP for pollination service came to 2.52 half-days of labour or lunch-time meals per week. This is equivalent to 131.04 half-days of labour or lunch-time meals per year. Thus, the monetized household WTP value for pollination service is US$ 87.80 per household per annum. The median annual household WTP value, which might be more appealing to the policy makers, since it's the amount many individuals are actually willing to pay for pollination service, is US$ 69.68. This household WTP value is seen as fair and not exaggerative considering the welfare of the people in the region. It can be compared with the gross net income (gross margins) derived from farming of major crops (maize, tea, and sugar cane) in the research area which has been estimated as US$ 395.91 per annum (Guthiga et al. 2006). If our sampling frame of 19,972 households that live within 10 km from the edge of the Kakamega forest is considered, the aggregated annual economic value of pollination service is US$ 1.75 million (or US$ 1.39 million considering the median value). The high aggregate WTP value indicates that pollination has high economic value that can support arguments and initiatives geared towards pollinator conservation in the areas surrounding Kakamega forest. It also shows that the community feels pollination has a role to play in their well-being and livelihood. The results further indicate that people are highly motivated in tapping the benefits of pollination service.

Determinants of WTP

With the highly positive response to the principal WTP question, it is necessary to assess the factors that might have contributed to this result. In addition, it is important to analyse factors favouring either of the payment vehicles chosen by the respondents. Thus, different factors that influenced the respondents' mean WTP value and that corresponded to a particular payment vehicle were determined by use of selected demographic variables in three different Tobit regression analyses (Table 5.3).

The results of the Tobit model show that only education and off-farm income significantly determined the magnitude of the sample WTP. According to our results, respondents with more years of schooling were likely to pay higher WTP values for the pollination service, while those with off-farm sources of income were likely to

Table 5.3: Coefficients and standard errors from Tobit regression analyses where WTP value for pollination service was regressed on socio-economic variables

Explanatory variable	Labour WTP	Meal WTP	average WTP
Constant	1.7045	−2.0360	2.7265
	(0.6475)***	(1.2207)*	(0.3774) ***
RELIGION	0.3043	−0.2744	−0.0584
	(0.2723)	(0.5045)	(0.1525)
PERC_POL	0.0602	−0.1752	−0.0254
	(0.3116)	(0.5523)	(0.1721)
DIST_FOR	0.1653	−0.4496	−0002
	(0.2588)	(0.4800)	(0.1448)
GENDER	0.7164	−1.2174	0.0302
	(0.2762)***	(0.5154)**	(0.1540)
MARITAL	−0.1603	0.9380	0.1967
	(0.3198)	(0.6165)	(0.1793)
AGE	−0.0223	0.0265	−0.0059
	(0.0089)**	(0.0158)*	(0.0049)
EDU	−0.0144	0.0768	0.0327
	(0.0334)	(0.0598)	(0.0187)*
HH_S	0.1301	−0.2838	−0.0344
	(0.0568)**	(0.1034)***	(0.0309)
TOT_EXP	−0.4783E−04	0.7278E−04	0.6389E−05
	(0.2541E−04)*	(0.3461E−4)**	(0.1176E−04)
OFF_INC	−0.5079	0.4591	−0.2412
	(0.2609)*	(0.4815)	(0.1459)*

Notes: Figures are coefficients, with standard errors in parentheses. Labour WTP and meal WTP refer to WTP generated using half-days of labour and lunch-time meals respectively.*, **, ***; significant at 10, 5; and 1 per cent level; WTP_Labour: Log likelihood= −593.4472, p 0.000; WTP_Meal: Log likelihood= −425.0300, p 0.000; average WTP: Log likelihood= −571.1626, p 0.000; N= 345; 6 observations left censored at 0.
Source: Authors.

pay less. Our findings agree with other studies that have documented that an educated society may be willing to pay more for nature conservation in developing countries. For example, Hadker et al., (1997) working in India reported in his study that an increase of one more year in school would raise the WTP value required to conserve wildlife by 5 per cent. Likewise Whitehead (1992) showed that the number of years in school influenced the willingness of the respondents to pay for loggerhead sea turtle conservation in North

Carolina, USA. Heinen (1993) also observed positive correlation between the attitudes of the respondents towards preservation of nature and the years of education.

Although total household income (proxied by total expenditure) has the expected positive sign, it does not significantly influence the sample WTP. This may be because our study did not ask for WTP in monetary terms. Using other forms of payment vehicles thus makes our results differ partly (not fully) from those reported in several other CV studies as far as the role of income in influencing WTP is concerned (see, for example, Boyle and Bishop 1987; Bandara and Tisdell 2004). The level of household income, however, significantly influenced the amount paid in terms of labour and meals though in different directions. As the results indicate, an increase in household income would make the households' WTP more in terms of meals but less in terms of labour. Related to the level of income is the source of the income. Normally, respondents in the research area with off-farm income sources also earn higher incomes. Since the opportunity cost of labour is higher for households with off-farm income sources, such respondents are more likely to be willing to pay less in terms of working (providing labour) in the hypothetical pollination garden. From the perspective of the sample WTP, it is most likely that sample households with off-farm income sources tend to pay less for pollination because their interests and investments in farming are low.

The influence of the location of the household (distance) relative to the forest on the sample WTP value was not significant though the variable had the expected sign. Respondents belonging to households far from the forest were likely to pay more for bee conservation than those close to the forest. This could imply that households near the forest considered the contribution of the forest as a pollinator habitat. Their acceptance to pay for pollination management could also indirectly imply that they would be willing to conserve the Kakamega forest although this was not captured directly.

Other factors that had a significant influence on the magnitude of WTP generated through the two different payment vehicles were gender, household size, and age. However, they show different directions of influence and levels of significance. Male respondents were likely to pay more in terms of labour while their female counterparts are likely to pay more in terms of meals. Usually males in the study area have fewer household chores compared to their female counterparts, and this could be the reason behind their choice of labour. Females, however, spend most of their time in crop fields and perform nearly all the household chores, for example, laundry, meal preparation (Canwat 2007, unpublished thesis) and may not have spare time for extra work in the proposed pollination garden. Therefore, it would be easier for them to pay for the pollination service through provision of meals.

As hypothesized, respondents belonging to larger households were likely to pay more in form of labour and less in terms of lunch-time meals. This is mainly because there is an incomplete market for labour in the research area. Thus, compared to small-sized households, larger households may have spare labour and thus, it would be easier for them to provide labour than meals. Also, as expected a priori, members of older households may not be energetic enough to provide labour for the pollination garden. They are, therefore, likely to pay more in terms of meals but less in terms of labour.

CONCLUSIONS AND POLICY RECOMMENDATIONS

This study was conducted among households living around the Kakamega forest in Kenya not only to assess their perception of the importance of pollination but also to elicit their WTP for the improvement of this ecological service. It is clear that farmers were aware of the existence of pollination and understood its importance in crop production. Thus, policies to support their needs in terms of improving provision of pollination service in the agricultural landscapes of the research area are required. Such policies are likely to favour conservation of the Kakamega forest, which is a key reservoir for pollinators in the region.

The high positive response of the households to the WTP question implies that the hypothetical market presented to them was appealing. The valuation exercise done using the CVM has numerous implications for the conservation of pollinators in Kakamega. The magnitude of the WTP, in particular, shows that there is a strong economic argument for conserving pollinators in the research area. The economic value generated also reflects the value of pollination service from the perspective of the users (consumers, growers), something that could not have been accomplished through application of RP methods, which rely on assumptions based on observable market transactions. This is the strength of this valuation exercise in the context of developing countries where pollination markets are lacking.

The results of this study indicate that educating local households impacts positively on nature conservation and especially on the pollination service. Policy makers, government, and non-governmental agencies should, therefore, continue their efforts to support education, for example, by maintaining the free primary education system. Further, enhancement of farming activities as a source of income generating opportunities and activities for the rural population would also play an important role in nature conservation and in the provision of pollination services.

The study also tested two in-kind payment vehicles—provision of labour or meals—which were found to be more appealing than making monetary contributions to the hypothetical project. Thus projects designed to allow public participation in the conservation of pollinators are more likely to succeed if they allow in-kind payments. Although the results show that contribution of labour was preferred to lunch-time meals, it is difficult to rule out the application of the latter. This is because there are particular groups of households, for example, the higher income earners and the elderly, which are likely to provide larger payments when given the option to pay meals. This calls for an appropriate choice of the in-kind payment vehicle, since its effectiveness would depend on the group of local households being targeted for a particular pollination conservation initiative.

REFERENCES

Allport, R.C. and J.E. Epperson (2003), 'Willingness to Pay for Environmental Preservation by Ecotourism-linked Business: Evidence from the Caribbean Windward Islands', FS-03-01, University of Georgia, USA.

Bandara, R. and C. Tisdell (2004), 'The Net Benefit of Saving the Asian Elephant: A Policy and Contingent Valuation Study', *Ecological Economics*, Vol. 48, pp. 93–107.

Binger, Brian R., and Elizabeth Hoffmann (1988), *Microeconomics with calculus*, Glenview, Scott, Foresman and Company, Illinois.

BIOTA (Biodiversity Monitoring Transect Analysis in Africa) (2004), 'Biodiversity in conversion: The influence of fragmentation and disturbance of East African highland rainforest', Final report of Phase 1.

Boyle, K.J. and R.C. Bishop (1987), 'Valuing Wildlife in Benefit–cost Analyses: A Case Study Involving Endangered Species', *Water Resources Research*, Vol. 23, pp. 943–50.

Canwat, V. (2007), 'Economic Analysis of Labour Resource Allocation Within the Rural Economy of Kakamega District, Kenya', Masters Thesis, University of Bonn, Germany.

Carreck, N. and N. Williams (1998), 'The economic value of bees in the UK', *Bee World*, Vol. 79, No. 3, pp. 115–23.

Carson, R.T., N.E. Flores, and N.F. Meade (2001), 'Contingent Valuation: Controversies and Evidence', *Environmental and Resource Economics*, Vol. 19, No. 2, pp. 173–210.

CBD (Convention on Biological Diversity) (2001), *Agricultural biodiversity: Progress Report on the Implementation of the Programme of Work, Including Development of the International Pollinators Initiative*, Montreal, Canada.

Costanza, R., R. D'Arge, R. de Groot, S. Farber, M. Grasso, B. Hannon, K. Limburg, S. Naeem, R.V. O'Neill, J. Paruelo, R.G. Rifkin, O. Sutton, and M. van den Belt (1997), 'The Value of the World's Ecosystem and Natural Capital', *Nature*, Vol. 387, No. 6630, pp. 253–60.

de Groot, R.S., M.A. Wilson, and R.M.J. Boumans (2002), 'A Typology of the Classification, Description and Valuation of Ecosystem Functions, Goods and Services', *Ecological Economics*, Vol. 41, pp. 393–408.

Ellingson, L. and A. Seidl (2007), Comparative Analysis of Non-market Valuation Technique for the Eduardo Avaroa Reserve, Bolivia', *Ecological Economics*, Vol. 60, pp. 517–25.

FAO (Food and Agriculture Organization of the United Nations) (2000), 'Applications of the Contingent Valuation Method in Developing Countries: A survey', FAO Economic and Social Development Paper 146, FAO, Italy.

Freeman, A. Mayrick III (2003), 'The Measurement of Environmental Values and Resources: Theory and Methods', Resources for the Future, Washington D.C.

Gikungu, M.W. (2006), 'Bee Diversity and Some Aspects of their Ecological Interactions with Plants in a Successional Tropical Community', PhD Thesis, University of Bonn, Germany.

Gordon, J. and L. Davis, (2003), *Valuing Honeybee Pollination, Rural Industries Research and Development Corporation*, Publication no. 03/077, RIRDC, Australia.

Greiner, A., (1991), 'Nature Conservation and Local People at Kakamega Forest', ASA Project, Final report, Stuttgart, Germany.

Guthiga, P., K. Holm-Müller, and J. Mburu (2006), 'Cost-benefit Analysis of Different Management Approaches of Kakamega Forest in Kenya', Eighth annual BIOECON conference on economic analysis of ecology and biodiversity, 29–30 August, Cambridge Kings College, UK.

Hadker, N., S. Harma, A. David, and T.R. Muraleedharan (1997), 'Willingness to Pay for Borivli National Park: Evidence from a Contingent Valuation', *Ecological Economics*, Vol. 21, pp. 105–22.

Hein, L. and J. Mburu (2006), Methods of Economic Valuation of Pollination Services, FAO, Rome, Italy.

Heinen, J.T. (1993), 'Park–people relation in Kosi Tappu wildlife reserve, Nepal: A Socio-economic Analysis', *Environmental Conservation*, Vol. 20, pp. 25–34.

Jaetzold, R. and H. Schimdt, (1982). 'Farm Management Handbook of Kenya', Vol. II/A, West Kenya, MOALD, Kenya.

Kahneman, D. and J.L. Knetsch (1992), 'Valuing Public Goods: The Purchase of Moral Satisfaction', *Journal of Environmental Economics and Management*, Vol. 22, pp. 57–70.

KIFCON (Kenya Indigenous Forest Conservation Programme), (1994) 'Kakamega forest, The official Guide', Nairobi, Kenya.

Kevan, P.G. and T. Phillips (2001), The Economics of Pollinator's Declines: Assessing the Consequences', *Conservation Ecology*, Vol. 5, No. 1, p. 8 (*http://www.consecol.org/vol5/iss1/art8*).

McDonald, J.F. and R.A. Moffitt (1980), The Uses of Tobit analysis', *Review of Economics and Statistics*, Vol. 62, No. 2, pp. 318–21.

Mitchell, Robert Cameron and Richard T. Carson (1989), 'Using Surveys to Value Public Goods: The Contingent Valuation Method', John Hopkins University Press, Baltimore.

Olmstead, A.L. and D.B. Wooten (1987), 'Bee Pollination and Productivity Growth: The case of Alfalfa', *American Journal of Agricultural Economics*, Vol. 69, pp. 56–63.

OECD (Organisation for Economic Co-operation and Development) (2002), *Handbook of Biodiversity Valuation: A Guide for Policy Makers*, OECD, France.

Pearce, David W. and R. Kerry Turner (1990), *Economics of Natural Resources and the Environment*, Herefordshire, Harvester Wheatsheaf, UK.

Perman, Roger, Yue Ma, James Mcgilvray, and Michael Common (2003), *Natural Resource and environmental economics*, London, Pearson Education Limited, UK.

Sanford, M.T. (1995), 'Pollination Requirements of Vine Crops', in H. Hochmuth and N. Maynard (eds) Vegetable Crop Proceedings, Florida Agricultural Conference and Trade Show, pp. 36–8.

Southwick, E.E. and L.Jr. Southwick (1992), Estimating the Value of Honeybees (Hymenoptera: Apidae) as Agricultural Pollinators in the United States', *Economic Entomology*, Vol. 85, pp. 621–33.

Tanguay, M., W.L. Adamowicz, and P. Boxall (1995), An Economic Evaluation of Woodland Caribou Conservation Programmes in Northwestern Saskatchevan', Programme Report 95–01, Canada: University of Alberta.

Tisdell, C., H.S. Nantha, and C. Wilson (2007), Endangerment and likeability of wildlife species: How Important Are They for Payments Proposed for Conservation?', *Ecological Economics*, Vol. 60, pp. 627–33.

Tobin, J. (1958), 'Estimation of Relationships for Limited Dependent Variables', *Econometrica*, Vol. 26, pp.24–36.

Whitehead, J.C. (1992), 'Ex-ante Willingness to Pay with Supply and Demand Uncertainty: Implications for Valuing a Sea Turtle Protection Program', *Applied Economics*, Vol. 24, pp. 981–8.

6

A Pilot Study on Payment for Ecological and Environmental Services in Lashihai Nature Reserve, China*

CARLO GIUPPONI, ALESSANDRA GORIA,
ANIL MARKANDYA, AND ALESSANDRA SGOBBI

INTRODUCTION

Ecosystems, natural or managed by humans, provide a wide range of benefits and ecological and environmental services (EES). These benefits can take various forms: ecosystems may provide goods which are used for consumption or production, such as food or water (provision service); they may serve to regulate other ecosystems' activities, for instance, regulating flow or pollution diffusion (regulating services); they may also sustain underlying processes which maintain productive assets, such as ensuring nutrient cycles (supporting services); and finally they may provide recreational, spiritual, religious, and other non-material benefits (cultural services).

This chapter discusses the preliminary findings of a pilot study on PES in the Lashihai Nature Reserve (LNR) (Yunnan Province, China). The pilot study aims at identifying the main EES, and the most appropriate payment schemes that can provide incentives to landholders to maintain or generate the services identified.

The chapter is organized as follows. Based on a thorough review of the literature, the next section describes existing PES schemes, and provides an assessment of their effectiveness in maintaining key

*This report provides the preliminary findings of a pilot study on PES in the Yunnan province of China, conducted by Fondazione Eni Enrico Mattei (FEEM) in cooperation with Conservation International-China, under World Bank funding (Italian CTF030815). We would like to thank He Yi of Conservation International, Jian Xie of the World Bank, and Zhou Ting of China Agricultural University for their comments and contributions to this study.

EES (sub-section 'Assessing the Effectiveness of PES Schemes'). The main experiences with PES schemes in China are presented in sub-section 'Application of PES Schemes in China', while sub-section 'A Pilot Design of PES Schemes: LNR' introduces the pilot case study. The third section presents the methodology adopted in the case study area. The fourth and fifth sections discuss how the value of the EES and the costs of provision have been calculated, respectively. The sixth section discusses the different options for designing a PES scheme in the pilot area, while the seventh section concludes.

PES SCHEMES

Despite their importance, EES are often lost, especially when they are controlled by private interests: landholders, for example, have no incentives to preserve them, as the benefits are enjoyed by many people, while the costs of maintaining them are incurred only by landholders. In recent years, compensation to landholders for the services generated by their land has been advocated for, as an instrument to ensure that these services are maintained. PES seek to capture at least part of the benefits derived from environmental services (such as clean water) and to channel them to the landholders who generate them: PES provide landholders the right incentives to maintain a healthy ecosystem, they are a new source of income for landholders who can improve their livelihoods (Pagiola et al. 2005), and have the additional advantage of generating funds that can be used to finance conservation projects.[1]

The underlying rationale of PES schemes is relatively simple and appealing, yet implementing them may not be easy (Pagiola and Platais 2003). The design of the schemes must necessarily be tailored to the local situation, not only in terms of the service traded, but also taking into account the current institutional constraints, as well as the capacity (financial and human) of potential suppliers and beneficiaries of the service. A taxonomy of the different PES schemes is given in Table 6.1. Some comments on the relative merits of these schemes are also provided.

---

[1]Note that the principle behind the PES is one of the 'beneficiary pays'. This is in contrast to the 'polluter pays', which is enshrined in the environmental regulations of many countries. Ultimately, the choice of which principle to adopt is one of equity (who should really pay for environmental protection) and of practicality (is it feasible to make polluters pay).

Table 6.1:   Main types of PES schemes

Type of PES	Participants	Type of EES	Requirements
Voluntary contractual agreements (VCS)	Private to private Role of government limited to enforcement of property rights	High-value EES, related to private good Low cost of provision of EES Small scale	Clear and enforceable property rights Negotiable contracts Limited number of providers and beneficiaries
Public payment schemes (PS)	Government to private, government to government, or government to other organizations (for example, NGOs, community based organizations,...)	Public good, significant externalities involved High value of EES, but high cost of provision	Generation of funds for government (taxes, user fees,...) Transparent institutions Public participation
Trading schemes (TS)	Private to private, with government setting initial standards and allocation of rights	High value of EES, variable costs of provision, EES related to private good Services by different providers must be perfectly substitutable	Strong institutional setting, Strong monitoring and compliance mechanisms, Clear initial allocation of rights

Source: Authors.

From the theoretical point of view, there are three key elements for the effective design and implementation of PES schemes: first of all, it is necessary to define and quantify the value of the EES to be maintained or enhanced; all the stakeholders—potential beneficiaries and providers of the services, as well as institutions—need to be defined, and finally the appropriate institutional set up—including the payment level and mechanisms—must be established

Assessing the effectiveness of PES schemes

PES schemes are relatively recent, and, as stressed by Landell-Mills and Porras (2002) in their extensive review, very few studies undertake an objective and comprehensive assessment of the costs and benefits associated with such schemes, including their effectiveness in mitigating environmental deterioration. It is nonetheless possible to identify some general conditions, which are likely to increase the effectiveness and efficiency of this tool in achieving the desired

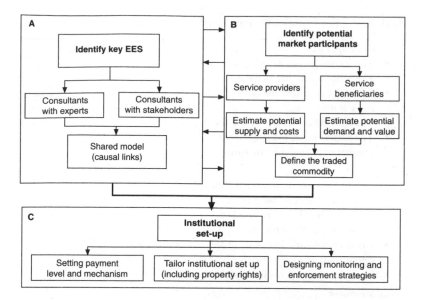

Fig. 6.1:   Implementing PES Schemes

environmental objective. These are synthesized in Table 6.2. PES schemes will be effective if the payments reach the providers of the services and motivate them to change (or not change) their land use practices. Since EES are a flow that will be maintained through the years, service providers must receive a stream of payments as long as they maintain the ecosystem; payments should be targeted and tailored, both to the level of the service provided and to the quantity demanded. Since targeted schemes entail higher transaction costs, however, a balance needs to be found. Furthermore, experience shows the need for a strong institutional and political commitment. If most of these conditions are satisfied, PES schemes can help ensure the maintenance of EES.

Application of PES schemes in China

The growing attention to PES schemes as instruments for environmental management is reflected in the recent strategies adopted in China, where the government has developed and implemented some of the largest public payment schemes for ecosystem services in the world. PES schemes have been experimented with since the 1980s, when the Ministry of Water Resources attempted to protect fragile

Table 6.2:   Key factors affecting the effectiveness of PES schemes

---

Type and value of EES
- The service provided is unique, scarce, and not easy to substitute
- The service to be transacted is clearly identified
- It is easy to quantify the value of the service to the beneficiaries, and the costs of provision
- The link between the service and the quality of the ecosystem providing it is clear
- The market is not too large (both in terms of geographical extension and number of potential participants)

PES design
- The PES scheme arrangement is flexible
- Transaction costs are minimized
- Compensation levels are based on the estimated value of the economic importance of the service to the beneficiaries
- Payments are sufficient to cover the cost of provision (including the opportunity cost of alternative land use)
- Compensation reaches both landowners and users

Social dimension
- Stakeholders can actively participate in decision making
- All the relevant stakeholders are consulted, including vulnerable and marginalized groups
- The distribution of costs and benefits is deemed acceptable.
- The PES scheme does not have adverse impacts on equity and poverty
- The beneficiaries are willing and able to pay
- The funds generated are invested in ecosystem maintenance

Institutional setting
- The political situation is stable
- Clear and uncontested assignment of property rights exists
- An adequate institutional framework is in place
- Compliance, land use change, and the provision of the service are closely monitored
- An enforcement strategy exists
- PES are managed in a fair and transparent way

Source: Authors.

watersheds through contracting out of land in sensitive areas to households, conditional upon appropriate management practices being adopted (Liu 2005). In fact, as early as 1991, market mechanisms for watershed management were introduced in Chinese legislation (Water and Soil Conservation Act), though with limited success.

The Sloping Land Conversion Programme (SLCP) was initiated in 1999 to restore natural ecosystems and mitigate the adverse impacts of agricultural practices carried out in previously forested areas or marginal land, such as flooding, sedimentation of reservoirs, and dust storms. The Government has already spent over RMB 50 billion[2] which has resulted in the enrolment of over 7 million ha of cropland (Xu et al. 2006), but financial constraints are forcing the slowdown of SLCP implementation (Scherr et al. 2006). Another major PES scheme is the Forest Ecosystem Compensation Fund (FECF), which targets the management of privately-owned standing forests, with the aim of compensating landowners for the ecosystem services their land provides and for the land and resource use restrictions the programme participation entails. This scheme currently covers 26 million ha in 11 provinces, and costs the Government about RMB 2 billion[3] annually, of which about 70 per cent goes to farmers for an average payment of $9/ha.

Neither of these schemes has achieved all of the targeted results, and there is now growing concern over their financial sustainability (IIED 2006; Xu et al. 2006; White and Martin 2002). According to Xu et al. (2006), the FECF and SLCP have made progress in the protection of forests—where logging bans, harvest reduction, and resource protection targets have largely been met. As a result, about a half of the logging, hauling, and processing assets (30 billion Yuan) in state-owned forest enterprises have been abandoned. To some this is a high price for the environmental benefits obtained. Further, both schemes rely heavily on state finances, which can be volatile. The lack of coordination among different, and often competing, institutions with management responsibilities is rarely cited as an obstacle, yet it hampers the efficient and effective implementation of PES schemes. Finally, Xu et al. (2006) claim that the projects have significant social impacts, which, however, have never been assessed. For instance, White and Martin (2002) highlight that the logging bans included in the SLCP and FECF, which originally targeted public forests, have in some provinces been extended arbitrarily to community forests.

There are many other smaller scale examples of PES tools in China, such as the water rights trading scheme between Yiwu and Dongyang cities in Zhejiang Province, the evolving framework of integrated watershed management and payments being developed between

[2]Approximately 6.54 billion US$ at current exchange rates (May 2007, 1RMB= 0.13 US $ (May 2007).
[3]Approximately 261.5 million US $ at current exchange rates (May 2007).

Beijing, Tianjin, and local governments in the upper watershed of the Miyun reservoir, or the experimental emissions-trading scheme conducted jointly with the Environmental Defence Fund and taking place in four provinces and three cities (Zhou 2005).

These examples of local and regional initiatives suggest that there exists significant interest and potential for the application of PES in China, even though till date such schemes have been in the more developed and rich provinces, such as Zhejiang, Fujian, and Guangdong. There are several other shortcomings of current PES mechanisms, notably: (i) the need to ensure that the schemes are financially sustainable, given the budgetary constraints faced by the Government (Scherr et al. 2006), (ii) the unclear nature of current land ownership and property rights over natural resources, and (iii) the lack of stakeholder participation in the design of PES systems.

So far, PES schemes in China have mostly been driven by government intervention, with little demand or supply forces behind their implementation. The situation is gradually changing, as people become more and more aware of the underlying ecological systems, beneficiaries are more willing to pay to maintain the service they enjoy, and service providers are more aware of their rights to cover the costs of service provision. Seemingly, the Government will continue to be heavily involved in the schemes but perhaps the country will move away from schemes where the Government is the sole buyer of the service to one where it will play a support role in creating and maintaining an enabling environment for private actors' transactions.

A pilot design of PES Schemes: LNR

The LNR was established in 1998 at the Yunnan provincial level, with the main purpose of protecting the Lashihai wetland, an area of special interest for migratory birds, listed under the RAMSAR Convention. The Lashihai Wetland is a unique plateau freshwater lake with marsh meadows, located between 2440 and 3100 meters above sea level at the headwaters of the Yangtze River in the Hengduan Mountains. It is an important migration passage, breeding ground, and wintering habitat of nearly 200 species (between early October and late April), among which are 76 species of wild geese and ducks, and protected wildlife such as the black-neck crane. As a biodiversity 'hotspot', Lashihai attracts 200 to 300 tourists daily, particularly for bird watching and horse-riding. Major protection measures include strict control (including some bans) on fishing, and hunting, but increasing unplanned tourism and agricultural activities are a potential threat for the lake's ecosystem.

There are four villages around the lake, with some 7500 people including fishermen and farmers. The latter grow mainly wheat, maize, potato, and rapeseed (Quan et al. 2002). Farmers who cultivate land around the lake suffer substantial yield losses because of the migratory birds, who feed on their crops. There is, therefore, a concern that farmers may harm the protected wildlife, unless adequate compensation is received, or alternative livelihood opportunities are offered.

Also, during the summer months (May to July), the Lashihai Lake supplies water to the Lijiang old town, named the 'Venice of China', a tourist destination where water canals provide a unique identity to the city. The old city, which relies on tourism development, is facing water scarcity and poor water quality. In relation to the application of PES schemes for water management in the Yulong province, the main concern at this stage is the water quality of the Lashihai Lake, as poor quality of the lake may harm the tourism industry in the old city of Lijiang during the peak summer months. Farmers with plots around the Lashihai Lake have the potential to reduce pressure on water quality by adopting more environmentally friendly agricultural practices.[4]

The LNR PES pilot study thus proposes to explore the potential usefulness of PES mechanisms to ensure that biodiversity services are maintained, and water quality for landscape use is restored.

METHODOLOGY

In this section, the methodology adopted for analysing the problem and identifying potential responses is described. First, the service providers and beneficiaries were clearly identified, as summarized in Table 6.3.

[4]Note that there is an implicit assumption here, namely that the status quo in terms of agricultural practices and, subsequently, water quality is accepted as 'legal'. That is, it is assumed that farmers in the LNR have the right to use their land as they wish, although in line with the general legislative framework of China. On the contrary, the people in the Lijiang old town do not have the right to clean water. This implicit assumption is supported by the views of the people interviewed during the course of the project, and the current stand of the local government.

Table 6.3: Summary of EES, service providers, and service beneficiaries

EES Service	Service providers	Service Beneficiaries
Improved water quality for landscape services	Farmers around the Lashihai Lake	Citizens of Lijiang Tourism industry—Lijiang old town Visitors to the old town
Maintenance of birds' biodiversity	Farmers around the Lashihai Lake	Tourism industry—LNR Visitors to the nature reserve Global benefits—biodiversity preservation

Note: Global benefits are traditionally not included in local PES schemes.
Source: Authors.

In order to identify possible leverage points for the maintenance of key ecosystem services provided by farmers in the LNR—that is, factors which, when externally modified, can lead to altering the behaviour of the system—a qualitative model of cause-effect relationships, linking human actions and the environmental system was developed and used for facilitating problem analysis and communication with various actors. The DPSIR approach was adopted as a reference conceptual framework (EEA 1999), because it allows for the integration of the various factors (quantified by specific indicators), having a role in the problem under scrutiny, their connection in context-dependent cause–effect chains, thus describing the relationship between the underlying causes and impacts on a system, and identifying and assessing current (or potential) policy responses to de-couple pressures and impacts. The DPSIR framework is based upon the consideration that one or more driving forces (D) cause one or more pressures (P) on the ecosystem, which in turn determine the state of the environment (S). Analysing changes of the state, impacts (I) can be assessed and evaluated, and responses (R) identified.

In the case of the LNR PES scheme, the DPSIR framework was used to conceptualize the underlying model linking farmers' activities to, on the one hand, the quality of water used for landscape services in Lijiang old town; and, on the other hand, protected bird populations in LNR.

The key indicators to monitor the state of the environment—and, therefore, the effectiveness of the PES scheme in maintaining the two EES—have been classified into four several categories:

1. water quality—both in lake (introducing among the routine monitoring indicators some parameters directly linked to agricultural activities, such as nitrates), and in Lijiang (visual impact, for example, algae concentration);
2. biodiversity—abundance of birds' population;
3. water availability, for example, hydrological water balance of the lake; and
4. ecological status of the wetland.

The state of the environment with respect to the two identified EES is determined by the interactions among several factors, the most important of which are agricultural practices (use of chemical inputs, tillage, irrigation systems) and illegal hunting of birds. In turn, these are driven by root causes, such as economic development (of tourism in Lijiang old town as well as LNR); government policies regulating agriculture and tourism development; and market demands for agricultural products.

Through the impact on the state of the environment, the driving forces push the system towards a state of unsustainability, both economic (reduced livelihood opportunities for farmers in the LNR, adverse impact on the tourism sector), and ecological (lake pollution, reduced birds' population).

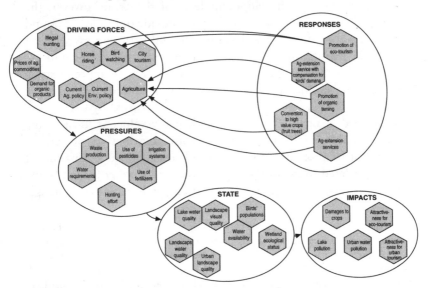

Fig. 6.2: A DPSIR conceptual model for human-environmental cause-effect chains in the Lashihai Nature Reserve

ESTIMATING THE VALUE OF THE EES

Assigning values to EES is one way for improving decision making for natural resource management, taking into account all the implications of different land use options. The economic value of an environmental service is generally measured in terms of how much beneficiaries are willing to pay for the commodity, net of the costs that suppliers incur in maintaining the EES. In the application of PES mechanisms, which are based on the creation of markets for transacting the EES to be maintained, enhanced, or restored, the values which need to be estimated are strongly related to the perception of people (in this case, of the service beneficiaries). In this context, 'value' is intended as the measure by which EES contribute to people's welfare, that is, it does not exist per se,[5] and refers to the change in people's welfare resulting from changes in the level of EES provision (in terms of its quantity or quality).

Several techniques are available to evaluate EES (see, for instance, Freeman 1993; Markandya et al. 2002). One could for instance assume that the revenues generated by eco-tourism activities around the LNR are a proxy for the value of the EES provided by farmers in terms of maintaining a healthy bird population. Given the current social and economic development level of the area, however, the value of the tourism industry is likely to provide only a lower bound for this EES, as there is potential for the industry to grow without necessarily hurting the Nature Reserve. Alternatively, one could use the expenditure for travelling to the Nature Reserve as approximating the individual estimates of biodiversity services provided by the reserve itself. In this case, however, the travel costs incurred by tourists are not deemed a suitable proxy for biodiversity value, as most of the tourists do not travel to the LNR as a final destination, but visit it because of its proximity to Lijiang old town. In the evaluation of the EES related to maintaining good quality of water for landscape use in Lijiang old town, it would be extremely difficult to disentangle

[5]Some may argue that ecosystems, such as wetlands, or biodiversity, have an intrinsic value, irrespective of whether humans enjoy it or not. According to this view, preserving environmental goods and services is a matter of moral obligation. In the perspective of PES mechanisms, however, only goods and services which can be traded in market-like transactions should be considered, whereas other tools and means are needed to ensure that the existence value of the natural environment is preserved.

from the value of the tourism industry—or from travel costs—that portion that can be attributed to water quality, rather, than, say, the quality of service, or architectural beauty. Our analysis thus relies upon the adoption of an SP approach, which is used to assign values to EES which do not have a real or estimated market/social value.

The questionnaire for the survey is the result of a joint effort between the research teams in Europe and in China. The first draft of the survey was tested on about 50 respondents, and changes were made to the structure and wording of the questionnaire to better reflect the needs of the Chinese language and the response and concerns of the focus group. The final survey was carried out on a total sample of 254 respondents—of which 50 were interviewed in the LNR, and the remaining 204 in Lijiang old town. The purpose of the survey was the elicitation of visitors' preferences with respect to the quality of their experience in either the old town or the LNR, and to assign an economic value to the identified environmental services. The questionnaire elicited respondents' view in relation to both the impact of poor water quality in the canals of Lijiang old town, and their real or hypothetical willingness to spend money to visit the LNR for bird watching or other eco-tourism activities.

Most of the respondents are Chinese nationals, about 67 per cent of the respondents are male, and over 50 per cent have a university degree. The distribution of average yearly income is a good approximation for the general distribution in China. About 27 per cent have visited Lijiang more than once—38 per cent of these have been to the city twice, while 13 per cent have visited it more than 10 times. Significantly, fewer tourists have visited the LNR more than once—87 per cent of the interviewees visited it for the first time. The mean length of the trip to Lijiang is 3.4 days[6]—with a maximum stay of 15 days.

Over 40 per cent of the respondents reported that they have not paid the government tourist tax of 40 RMB, that all tourists who stay overnight in Lijiang old town are supposed to pay. Yet, only 8 per cent of the respondents reported staying for one day only in Lijiang old town. This discrepancy could have two potential explanations: either most of the tourists do not sleep in the historical

---

[6]Note that several respondents answered very large numbers, which have been cleaned from the sample. The maximum stay for tourist purposes has been fixed at 15 days.

centre of Lijiang, to avoid paying the tax; or, as it has been voiced several times, the hotels—which are also required to charge tourists the visiting tax—do not enforce this requirement consistently.

Expenditure for a trip to Lijiang is, on average, 1950 RMB[7] per trip. The majority of these expenses are for travelling. In fact, 47 per cent of the respondents travelled to Lijiang by plane—the town is far from major cities, and there is no railway station. On average, people are satisfied with the services offered by the city of Lijiang, but think that there is still room for improvement. Contrary to expectations, most of the people think that water quality is excellent. This result is somewhat in opposition to the perception of local authorities, who perceive water quality in the Lijiang canals as poor and thus, harmful to the tourism industry. Of the tourists returning to the city, only 9 per cent have in the past had bad experiences because of the poor quality of the water in the canals. Of these, foul smell and impaired vision are ranked similar.

The main reasons for visiting the LNR are bird watching (30 per cent) and horse riding (25 per cent), while 26 per cent of the respondents do both activities. Thus, the absence of birds would significantly, and negatively, affect the pleasure derived from a visit

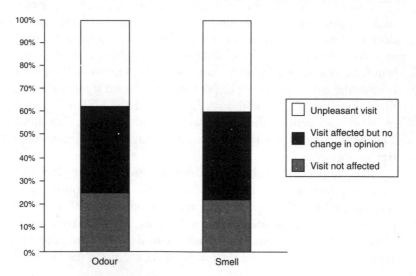

Fig. 6.3:   Respondents' opinion on landscape water quality

[7]Approximately 254 US$ at current exchange rates (May 2007).

to the reserve, with 61 per cent of the sample stating that they would derive significantly less pleasure from a visit to the reserve in the absence of birds, or if the opportunities for bird watching were significantly decreased. Indeed 63 per cent of the respondents stated they would not visit the Nature Reserve if there were no birds.

The average and median WTP for the protection of water quality and for biodiversity conservation are given in Table 6.4. The mean value of the former is lower than the latter, probably as a consequence of the majority of respondents not perceiving water quality as a problem. The median WTP is the same for both EES, but 78 respondents (30 per cent) stated that they would not pay to improve water quality.

Table 6.4:   WTP for the two EES

(RMB per annum)

	Mean WTP	Median WTP	Minimum	Maximum
WTP for landscape water quality	10.3	8	0	157
WTP for biodiversity in LNR	33.4	8	0	2500

Notes: The current exchange rate is 1RMB = 0.13 US$ (May 2007).
Source: Authors.
Based on this we approximate the use value of the two EES on the basis of the tourists visiting the two destinations.

Table 6.5:   The value of EES to service beneficiaries

(RMB per annum)

	Number of visitors	Value of the EES (mean)	Value of the EES (median)
WTP for biodiversity in LNR	65,000	2,171,000	520,000
WTP for landscape water quality	4,042,300	41,635,690	32,338,400

Source: Authors.

ESTIMATING THE COSTS OF PROVISION

The Nature Reserve Management office holds a registry of farmers' claims for compensation of the damages caused by birds.[8] The claims are verified by part-time employees of the Nature Reserve Management, who are local villagers, and are computed according

[8]The current exchange rate is 1RMB= 0.13 US$ (May 2007).

Table 6.6:   Estimated damages to farmers (selected birds species)

	2000	2001	2002	2003	2004	2005
Affected land (ha)	653	747	717	693	640	680
Estimated economic loss (RMB)	1,637,000	2,096,800	1,852,800	1,742,900	1,704,600	2,039,600

Source: LNR—Experts estimates.

to the surface area of the plot damaged by birds, an estimate of the yield lost to birds, and market prices for the crop. The estimated losses are in the range of 2,000,000 RMB per year,[9] yet compensation to farmers for the biodiversity service they provide, set at about 800,000 RMB per year,[10] falls short of this amount.[11] There is, therefore, a concern that farmers are harming the protected wildlife, and this may worsen unless adequate compensation is received, or alternative livelihood opportunities are offered.

As far as the second environmental service is concerned (that is, the quality of water in Lijiang), it is much more difficult to establish a clear, quantitative link between agricultural activities and the resulting improvement in water quality. A quantitative mathematical model would allow one to investigate these phenomena, but it would require accurate data on: (i) fertilizers and pesticides applied, (ii) on the observed water quality, (iii) on climatic patterns, and (iv) on soil type and land use. Most of this information is not available for the pilot area.

There are some spot observations and measurements which were undertaken by CI-China, which indicate pollution levels in the water bodies. For example, the concentration of the insecticide dimethoate was measured in water samples collected from both the Lashihai Lake and the canals in Lijiang old town. The tests show a concentration at the inlet of the Lashihai Lake of 79.7 ng/l, but concentrations increase in Lijiang's canals (247 ng/l in the middle of the town, and 365 ng/l in the lower part of the town).

[9]In particular, the main bird species that feed on farmland are: the bar-headed goose; the common and black-necked cranes; and the ruddy shelduck (Conservation International China).

[10]Approximately 261.000 US $ at current exchange rates (May 2007). Source: CI-China.

[11]Approximately 104.000 US $ at current exchange rates (May 2007). Source: CI-China.

Even though these concentrations are still low relative to most standards (20–50 mg/l), the data show an increasing deterioration of water quality in the city's canals. Furthermore, the CI-China team collected empirical evidence of pesticide use, such as 44 empty bottles of pesticides—some of which are forbidden in China—in the Lashihai wetland area.

From the above, it is clear that data limitations constrain the analysis of pollution phenomena to a qualitative assessment of the scale of the problem, according to the opinions of local policy makers. Therefore, it is clearly impossible to assess the likely consequences of potential strategies to reduce chemical inputs, in terms of (reduced) yield and (increased) costs. However, as we shall see, this gap does not invalidate the exercise, because there is still scope for improving water quality without farmers having to suffer an income loss.

DESIGNING THE PES SCHEME

The last step in the problem is the assessment of the different designs and targets of interventions of potential PES options (that is, the strategies which PES could adopt, sectoral targets) in terms of their impact on the variables of interest, namely water quality in the Lijiang old town canals in the summer and the affected birds' population. A preliminary list of implementation costs and expected benefits elements for each potential PES strategy is summarized in Table 6.7.

Agriculture is one of the main activities on which people living in the LNR rely. Thus, given the importance of agriculture and the fact that agricultural activities are considered to be responsible for deterioration of the water quality, a more detailed analysis has been carried out for this option. In particular, the following alternative agricultural practices were considered, in addition to the status quo:
1. traditional crops and fruit trees (D.I);
2. traditional crops and livestock (D.II);
3. traditional crops, livestock, and fruit trees (D.III).

For each of the options, a cost–benefit analysis was carried out,[12] and the results are summarized in Table 6.8. The values reported are the net present costs and benefits of alternative land use options over a time period of 15 years.

[12]Meeting with officials of the LNR Management Bureau, Yulong County Forestry Bureau, LNR, Yulong, China, 17 July 2006.

Table 6.7: Qualitative assessment of alternative PES options in the LNR

PES Option	Expected impact on income of farmers	Expected impact on water quality	Expected impact on birds' populations	Expected impact on visits to LNR	Direct costs of implementation	Indirect costs of implementation	Expected benefits to farmers	Expected benefits to tourism industry
Do nothing	Stable	Continued deterioration	Stable or declining (poaching)	Stable or declining (less bird-watching opportunities)	No additional costs	Negative externalities on birds' population and tourism in LNR		Decreased income in summer months—smell and visual impacts of poor water quality.
Alternative A: Agricultural extension Services	Possibly increase	Improvement	Stable or declining (poaching)	Stable or declining (less bird-watching opportunities)	Cost of extension services		Since current use of pesticides/fertilizers is excessive, farmers would benefit from lower input costs, without loosing in terms of crop yield.	Expect an improvement in water quality, hence no damage to the tourism industry in Lijiang old town.
Alternative B: Agricultural extension services with compensation	Possibly increase	Improvement	Stable or increasing	Stable or increasing (increase in birds' population)	Cost of extension services Cost of compensation		Expect that current use of pesticides and fertilizers is excessive. Hence, farmers would	Expect an improvement in water quality, hence no damage to the tourism industry in

(contd...)

Table 6.7 (contd...)

PES Option	Expected impact on income of farmers	Expected impact on water quality	Expected impact on birds' populations	Expected impact on visits to LNR	Direct costs of implementation	Indirect costs of implementation	Expected benefits to farmers	Expected benefits to tourism industry
for birds' damage					(possibly higher than current level)		benefits from lower input costs, without loosing in terms of crop yield. Furthermore, they would be fully compensated for damages caused by birds.	Lijiang old town. Expect a stop to illegal bird hunting, hence benefits in terms of increased birds population.
Alternative C Promotion of organic farming	? Depends on market demand	Improvement	Possibly decreasing (less food sources)	Possibly decreasing (less bird-watching opportunities; visual impact if greenhouses are used)	Initial cost of setting up organic farming structures, and of encouraging farmers to change. Cost of marketing	Possible negative impact on birds' population, as less food sources.	Expect organic farming is not economically viable, at least in the short term. Would need to subsidize the activity for one to two years.	Tourism in Lijiang old town would benefit from improved water quality, but tourism to LNR may suffer from visual impairment if organic farming is carried out in greenhouses, and a

(contd...)

Table 6.7 (contd...)

PES Option	Expected impact on income of farmers	Expected impact on water quality	Expected impact on birds' populations	Expected impact on visits to LNR	Direct costs of implementation	Indirect costs of implementation	Expected benefits to farmers	Expected benefits to tourism industry
					strategies and promotion of products.			possible decrease in birds' population
Alternative. D Conversion to high value crops (for example, fruit trees)	?	?	Possibly decreasing, depending on ecological niche of birds	Possibly decreasing, if birds' population decreases	?	?	?	?
Alternative E Promotion of eco-tourism	Possibly higher, but decreasing marginal returns	Possibly improving in the short term, but possible deterioration in the long term	Possibly declining, if fields are converted (less food sources for birds)	Possibly increasing. In the long term, potential loss of bird-watching opportunities	Cost of promoting LNR as a tourist destination locally and internationally. Capacity building cost—conversion of farmers to entrepreneurs.	Increase pressure on sanitation, possibly deterioration of water quality.	Expect this option not to be viable for all, unless cooperatives can be formed, and the site be promoted to attract an increasing number of tourists.	Benefits to both sites.

Source: Authors.

This preliminary analysis suggests that the most profitable alternative cropping patterns for the area are crops and fruit trees, either in isolation, or together with livestock. A mixed system of crops and fruits is already practised in some areas around the LNR, and can be further encouraged.[13] To increase farmers' profits, one could also encourage the introduction of livestock husbandry, which can reduce the dependence on chemical fertilizers. Further, in the last scenario, fodder crops intercropped under fruits may help to reduce rearing costs, further increasing the attractiveness of this development scenario for individual farmers.

The main purpose of changing the agricultural practices of farmers around the lake, however, is not to increase their profits, but to decrease the impact of their activities on the water quality of the lake, and, in turn, on the tourist city of Lijiang. It is thus necessary to understand the status quo in terms of agricultural yields, use of chemical inputs, and water quality in the lake.

According to expert opinion,[14] differentiating the types of crops cultivated in the area with high value fruit trees is expected to reduce the input of chemical fertilizers used around the LNR, while at the same time increasing farmers' revenues (see Table 6.9). Thus it serves as a 'win-win' alternative.

While this finding is encouraging, it needs to be qualified. Further specification of data and result is necessary to confirm the results. Further under the assumptions of the model, in scenario DI. Crops

Table 6.8:   CBA of alternative agricultural practices

Option	Costs (RMB/mu)	Benefits (RMB/mu)	Benefit–cost ratio
D0. Baseline case (crop only)	15390.39	12674.73	0.823549
DI. Crops and fruit trees	10815.37	13619.5	1.259272
DII. Crops and livestock	21295.32	18219.93	0.855584
DIII. Crops and livestock and fruit trees	12265.14	19164.7	1.562534

Source: Authors.

[13]The analysis was carried out by Zuo Ting, Deputy Dean and Professor in Development Studies, College of Humanities and Development, China Agricultural University.

[14]Some efforts in this direction are already in progress in the area, for instance by the Yunnan Centre for Biodiversity and Indigenous Knowledge.
www.cbik.org/cbik-cn/cbik/our_work/download/sustainable-2004 Annual Report.pdf

Table 6.9:   Input requirements of alternative land use options
(chemical fertilizers)

Option	Total Fertilizers (Kg/mu*) (average annual input over a 15 years period)	Variations with respect to the status quo
D0. Baseline case (crop only)	65.62	0 per cent
DI. Crops and fruit trees	61.26	−7 per cent
DII. Crops and livestock	47.30	−28 per cent
DIII. Crops and livestock and fruit trees	46.02	−30 per cent

Notes: *1 ha = 15 mu
Source: Authors.

and Fruits, the use of chemical fertilizers may increase over time to
maintain a stable yield, while in scenario DIII. Crops and Livestock
and Fruits, livestock manure will be used to make up for the lower
input chemical fertilizers; thus strengthening the attractiveness of
this farming option in terms of improved profitability, but at the
same time mitigating the positive impacts in terms of reduced
fertilizer use.

Even if the levels of fertilizer decline as a result of the shift in land
use, what impact will this have on water quality in the lake? In general,
the linkages between water quality and agricultural practices are very
difficult to assess, given the strong interdependencies among different
factors, such as climatic factors, temperature, land types and slopes,
biotic factors in the water body, and the overlapping effects of different
sources of pollution (that is, for the nutrient those related to civil
settlements). For the same reasons, extrapolations or analogies from
the international literature are not feasible, and the exercise would
not be helpful in policy making, as it could lead to decisions based
on misleading or outright wrong estimates.

The net impact on lake water quality cannot, therefore, be
assessed in a quantitative manner with the available information,
but one can nonetheless draw some general, qualitative, conclusions.

First, water quality data show that turbidity of the lake water is
relatively high, indicating potential problems with respect to soil
erosion, along with connected problems related to suspended nutrients
(phosphorus in particular). This would indicate the need to improve
land use practices in general, not only agriculture, but also the
management of forest and grazing, to control soil loss. Second,

available data show that current agricultural practices could be highly inefficient in terms of fertilizer use.[15] If confirmed with more robust information, the need emerges for a remarkable re-orientation of current practices, through an adequate extension effort. Third, the net impact of changes in agricultural practices cannot be easily determined: reductions in chemical fertilizers could be (more than) compensated by manure, in case of increasing livestock density. For this reason, and in general because of the complexity of the phenomena involved, the consequences of changing agricultural systems on water quality will remain uncertain.

Finally, an increase in fruit trees at the expenses of more traditional crops may have adverse impacts on the birds' population, depending on the ecological niches of birds. Further research is also needed in this direction, before a substantial change in cropping patterns can be encouraged.

In the light of the above considerations, PES schemes for the promotion of improvements in water quality through agriculture should focus on income-neutral or income-improving changes in farming practices, but ones that also support a shift towards soil conservation practices and more efficient use of agro-chemicals.

Promoting 'soft' changes which are income-neutral or income-improving, the estimation of the costs of service provision to the farmer is relatively less important—what is important in this case is the cost of implementing 'soft' measures (extension service) relative to the value that beneficiaries attach to the service.

Extension services

Inducing farmers to adopt more environmentally sound practices is likely to be cost-effective, as the local use of chemicals in agriculture appears to be quite above the average in China, indicating that farmers are not on the production possibility frontier. Integrated strategies targeting both inputs and soil management (for example, conservation practices and cultivation of less demanding crops) could be implemented, through carefully designed capacity building campaigns and extensions services. More stringent monitoring of farmers' activities would ensure the implementation of those practices, but also the

[15]Report by Zuo Ting, Deputy Dean and Professor in Development Studies, College of Humanities and Development, China Agricultural University.

enforcement of Chinese legislation on pesticides and fertilizers use in agriculture, which is already well developed.

Lower inputs and improved land use would be likely to lead to mitigation of environmental impacts, and eventually to an improvement in the water quality of the landscape water used in Lijiang old town. Thus, less traditional PES schemes could be implemented in this context: instead of direct (cash) payments to farmers for the EES deriving from their activities, capacity building and extension services could be provided. Funds to carry out these activities could still be sought from the service beneficiaries, in ways to be defined by the relevant ministries.

Organic farming

Organic farming—which, by definition, forbids the use of chemical inputs—could lead to significant improvements in water quality of the lake and thus of landscape water for Lijiang old town, and a reduction in water requirements for irrigation, lessening the conflicts over water sharing now and in the future.

The current organic farming practices promoted in the area entail cultivation in greenhouses. One must, therefore, consider that farmland around the Lashihai Lake is a prime source of food for important migratory bird species, and is located within a nature reserve of significant scenic beauty. The impacts that this shift in agricultural practices could have on both scenic beauty and the provision of birds' biodiversity needs to be further explored, to ensure that no negative effects are observed.

Therefore, for organic farming to be a feasible option one would need to explore the possibility of adopting practices that differ substantially from the observed trend in Yunnan, namely, without the use of greenhouses. Moreover, the need to subsidize farmers in the transition period (approximately two years), and to provide market opportunity, should be adequately considered.

Whether consumers' trust in Chinese organic produces can be built up or not is, at this stage of the study, a matter of speculation. According to the Organic Consumers Association,[16] organic food

---

[16]Data for the region show levels of application of N and P that are 12 to 20 times China's average, and application of K that is nearly three times the average. At the same time the yields in the region are lower than the national averages. The data are not entirely reliable but may be indicative of serious overapplication of fertilizers.

is set to grow substantially in China, and organic production in the country has been increasing by 30 per cent annually, with exports growing by 50 per cent in recent years (China Green Food Development Centre). Several sources indicate that Western buyers of organic products are increasingly turning to China for organic products.[17] Therefore, there seem to be good prospects for this initiative. Contract farming and the setting up cooperatives to exploit economies of scale may be a successful strategy.[18]

Institutional set-up and payment levels

As important as the types of interventions are questions relating to the institutional set-up which should be used for the PES mechanism. In particular, the authority needs to decide how the funds are channeled to the farmers who choose to participate voluntarily in the scheme, and how the needed financial resources are to be raised from service beneficiaries. These issues are discussed here.

Encouraging farmers' participation in the scheme

Two main lessons can be drawn from the review of best practices experiences that can be successfully applied in the context of the pilot study.

First, a scheme is likely to increase the chances of improving landscape water quality if compensation is conditional upon farmers adopting agricultural practices which require lower inputs and conservation practices. Ideally, such changes should generate increased incomes for the farmers in the long run anyway, thus reducing the PES to a transition programme.

Second, to maximize benefits for the environment, but also to promote the long term development of the region, the PES scheme could have an additional component, and could encourage selected farmers to adopt organic farming practices—perhaps without the use of greenhouses to avoid negative impacts on the birds' population and on the landscape of the nature reserve. It is probably desirable to limit participation in the organic farming component of the PES scheme to those farmers with land in

[17] http://www.organicconsumers.org/organic/china100804.cfm
[18] See, for instance, http://www.foodnavigator.com/news/news-ng.asp?n=59043-western-food-makers

sensitive areas—that is, right on the lake shore, or on land with a significant slope, where run-off is more problematic. Limiting participation in the scheme to only the critical farmers will help to keep the implementation costs low, but will require a specific survey to identify sensitive areas.

Charging service beneficiaries

The financial sustainability of the PES scheme is of paramount importance. Moreover, the possibility to raise the required funds from the beneficiaries is one of the appeals of this MBI in the face of considerable budget constraints for environmental protection and conservation. The review of the past experiences in China has once again highlighted how financial constraints are hampering the implementation of existing PES schemes.

First of all, and in line with the attempt to reduce, as much as possible, transaction and implementation costs, the payment vehicle should be easy to implement. Second are the targeted beneficiaries: on the one hand, the visitors of the LNR and, on the other hand, tourists to Lijiang old town.

Existing institutions and payment vehicles should be used, to reduce implementation costs while at the same time ensuring that local actors are already familiar with the institutions managing the scheme. Thus, it is suggested that the visitors' fee to the old town of Lijiang be used as a payment vehicle for collecting the revenue necessary to fund extension services and, in part, organic farming.

Beneficiaries of birds' biodiversity maintenance, at the local level, are mostly the visitors to the LNR. In this case, the easier strategy would be to introduce an entrance fee to the reserve. The fee should be set at an appropriate level (reflecting the WTP as given in Table 6.4) so as not to discourage tourist development of the site, which is just picking up.

Alternatively, a fee could be introduced on the eco-tourism activities: but this would probably harm local communities by lowering the profits from tourism. One of the main potential shortcomings of PES schemes is the risk of negatively impacting on wealth distribution: considering the development reality of the LNR, it therefore does not seem advisable to introduce a 'user fee' on eco-tourism activities to raise funds to compensate farmers for the maintenance of birds' biodiversity.

Setting payment levels

One of the most difficult exercises for the effective and efficient implementation of PES schemes is the definition of appropriate payments levels. The funds that the farmers receive should be sufficient to completely cover their true cost of providing the service and overcome their possible reluctance to change and be exposed to new bureaucratic burdens.

Without a quantitative model linking farmers' behaviour to different levels of water quality, it is not possible to have an estimate of the cost of service provision. However, given the current situation of the agricultural sector in the area, extension services are deemed a win-win solution to improve farmers' income as well as water quality. The additional fee imposed on the tourists of Lijiang should, therefore, be sufficient to cover the cost of providing agricultural extension services to all the farmers who wish to enroll in the scheme—the largest number possible. An approximation for the costs of extension services is summarized in Table 6.9, including the subsidies for organic farming.

To the above costs, one would need to add sufficient funds for marketing organic products, to ensure that farmers' reliance on state subsidies decreases over the years. This could be done by providing funds to cover networking costs to link the initiative to other initiatives in China (such as the BioFach initiative to promote organic farmers among smallholders as a means to reduce poverty),[19] with the funds managed by the Municipal Agricultural Bureau.

To cover the costs of provision, it is suggested that the local government implements a dual-system, through which international visitors will bear a higher burden of the additional fee. In general, demand is less affected by changes in price of the commodity, and the WTP is higher for international tourists than for domestic ones. The entrance fee for domestic visitors can be increased by 1 per cent, from 40RMB to 40.4 RMB, while the entrance fee for international visitors can be increased by 5 per cent, from 40 RMB

---

[19]It is also claimed that water use in organic farms using greenhouses is much lower than on ordinary farms (up to 80 per cent less). If correct, an expansion of such farms would help to improve water flow rates in Lashihai Lake and thus in the town's canals.

http://www.biofach-china.com/main/eefmgjhs/eefmtw9y/page.html

to 44 RMB. It is important that visitors are informed about the use of the additional fee: this awareness campaign strategy is likely to lower opposition to the increase. The estimated funds that would be generated by this increase are reported in Table 6.10, and would be sufficient to cover the costs of extension services, organic farming, and some funds would be available for marketing and networking activities.

As the number of tourists visiting the city is not known with certainty, it may be the case that the funds raised are above the needs for the year. In this case, the additional resources could be channeled to a revolving fund, which can serve as buffer for unexpected expenditure, to hedge the risks in commodity price volatility. The funds would need to be managed jointly by all the institutions involved.

Based on the estimates of damages caused by birds' population, and on the number of visitors entering the LNR every year, it is possible to estimate an entrance fee that would cover fully the cost

Table 6.10: Estimated cost of extension services

					(RMB per annum)
	Number	Units	Unit cost (US$)	Total cost (US$)	Total cost (RMB)
ES Staff	1	7 villages	2400 (salary)	16,800	132,806
Subsidies for organic farming*	1	1,487 ha	50	74,337	587,642

Notes: *upper bound—considers a subsidy of 50US$/ha for all the cultivated land around the lake, not only the critical land. This may only be needed for one to two years.
Source: Authors.

Table 6.11: Suggested increase in Lijiang visitors' fee—water only

	Domestic	International	Domestic + International
Increase	1 per cent	5 per cent	
RMB	0.4	2	
Number of paying visitors	2,315,700	109,680	
Funds generated (RMB/year)	926,280	219,360	1,145,640
Funds generated (US$/year) (US$/Year)	117,174	27,749	144,923

Source: Authors.

Table 6.12:   Suggested entrance fee to LNR—dual system

		Entrance fee RMB	Entrance fee US$
Number of domestic tourists	50,000	8	1
Number of international tourists	15,000	40	5
Total Revenue		1,000,000	126,500

Source: Authors.

of provision of the 'biodiversity service'. However, the level of a uniform fee sufficient to cover the cost of provision is well above the median WTP of 8RMB (approximately 1US$). Setting the entrance fee at such a high level may significantly disrupt the development of the tourism industry. As in the previous case, it is therefore suggested that a differentiated fee is charged to local and international tourists, to take into account the lower median WTP and the fact that, on average, international tourists are willing to contribute more towards the conservation of biodiversity. If the impact on visitors is not found to be significant, the entrance fee can be raised in the future.

In the short term, the funds raised through the entrance fee at a lower level will not be sufficient to fully compensate farmers for the damages caused by birds—with an estimated shortage of RMB 845,613 per year (approximately 110,000 US$). This is more or less the compensation currently disbursed by the LNR Management Bureau: the PES scheme could therefore help, in the short run, to bridge the gap between the observed damages and the compensated damages, thus reducing the incentives for farmers to harm the birds' populations, and facilitating the work of the LNR Management Bureau. In the medium to long run, additional funds could be raised through charging a higher entrance fee, once the impacts on the tourism industry have been assessed. As in the case of the funds for water services, it is important to ensure that tourists are informed about the use of the entrance fee, that is, that the purpose is to compensate farmers for the service they provide in terms of maintaining birds' biodiversity.

Institutional set-up

Currently, the visitors' tax to the city is managed by the Lijiang Municipal Tourism Bureau (MTB): the additional fee meant to cover the cost of agricultural extension services—and, if possible, a shift to organic production for a limited number of farmers—collected by MTB would need to be transferred to the Lijiang Municipal Agricultural Bureau (MAB), which is currently in charge of agricultural extension services. A campaign to train farmers into low impact agriculture, or organic farming, can also be funded through the additional fee for visiting Lijiang old town.

The LNR is managed by a local management office depending on the Yulong County Forestry Bureau (LNR-FB). The compensation for biodiversity service provision should therefore be managed by the LNR-FB. The current compensation that farmers receive under the Regulation on the Compensation for Personal Injuries and Property Loss caused by the key protected terrestrial wild animals in Yunnan Province (1998) is not sufficient to fully cover the yield losses.

It is important to point out that, as several government institutions would be involved in the implementation of the PES scheme, there is a strong need for coordination among them: first of all, compensation for birds' damage should be conditional upon farmers adopting environmental-friendly agricultural practices that minimize both chemical input and soil erosion. Close cooperation between the LNR-FB and the MAB is, therefore, a pre-requisite for the functioning of the scheme. In particular, extension services need to certify both farmers' participation and compliance with good agricultural practices.

To facilitate institutional interaction, a common database of participating farmers should be kept in the LNR-FB premises, where farmers' performances can also be recorded. The database will also record farmers' compensatory claims for lost crops to birds, and the damage assessment and verification carried out by either the agricultural extension workers or the LNR-FB employees. Clear mechanisms for transferring the additional funds received by the MTB to the MAB should also be set out, to ensure transparency and accountability.

Monitoring compliance

For the scheme to be effective, some strategies must be in place to ensure compliance of the participating actors: farmers and those supposed to manage the collection of fees. At a minimum, the MTB must ensure that the fee to visit the old town of Lijiang is paid by the visitor and, in case, that the dual system is properly applied. The MAB, instead should receive part of the resources collected for setting up extension and capacity building activities, together with the monitoring farmers' compliance with the schemes. It is also of paramount importance that the Environmental Protection Bureau begin routine monitoring of environmental indicators, in order to provide adequate ex ante, in itinere, and ex post quantitative information about the state of the environment and the effectiveness of the PES schemes. The LNR-FB will remain responsible for monitoring and assessing the damages suffered by farmers and for compensating those who are eligible (that is, the farmers who have adopted good agricultural practices).

CONCLUSIONS

Past experience in China as also international experience seem to indicate that setting up PES schemes for watershed services is fairly straightforward, due to the relative ease in identifying beneficiaries and providers, although some difficulties remain in quantifying the linkages between land management practices and outcomes. In the case of China, these types of PES schemes will not only help to improve protection of upper watershed services in face of weak enforcement of existing regulations, but can also serve as a useful platform to resolve the conflicts surrounding these services (Scherr et al. 2006).

In China, the private sector represents a potentially large source of financial resources to pay for ecosystem services through, for example, offset for pollution credits. There are already various examples in the country, such as the State Forestry Administration's Forest and Vegetation Restoration Fee, or the various fees levied by local governments for ecological damages caused from construction and engineering projects. It is now necessary to move to the next stage, in which PES schemes—and design processes—have to be standardized across the country, and the financial flow generated is invested in ecosystem restoration or protection activities.

This chapter has explored, through a mixture of qualitative and quantitative approaches, the potentials for applying PES schemes to preserve two key ecosystem services provided by farmers surrounding the Lashihai Lake, in the LNR: the maintenance/restoration of good water quality for landscape use in Lijiang old town and the maintenance of birds' biodiversity in the Nature Reserve.

The preliminary analysis of the two EES and their provision indicates that there is a substantial WTP on behalf of the service beneficiaries for maintaining the EES. In the case of birds' biodiversity, even though our preliminary results indicate that there is scope for establishing a PES scheme, the funds collected may not, in the short term, cover the full costs of compensation. In the case of water quality, the linkage between farmers' activities and service provision is still not clear. The analysis of agricultural practices does leave several opportunities open, as it is clear that farming activities are not efficient at the moment. There is, therefore, scope for improving water quality through 'soft' measures, such as extension services to improve farmers' practices, both in terms of yield, and to reduce soil erosion, another problem potentially present in the area, given the observed turbidity of lake water.

Importantly, this analysis seems to indicate that the two services must be addressed together in a bundled manner: this approach does not only allow for the exploitation and enhancement of potential synergies, but also the identification and avoidance of perverse effects that interventions to preserve one service may have on the other, such as the case of the promotion of organic farming.

The reasons for the increase in the visitors' fee to Lijiang old town, or the introduction of an entrance fee to the LNR, should be made explicit with adequate awareness campaigns with both the tourists and local citizens, to ensure that the value of the EES provided by the farmers in the LNR is recognized, thus reducing resistance to the introduction and maintenance of the PES.

Four key general lessons can be drawn from this study:
1. Whenever possible, it is advisable to use existing institutions and payment vehicles.
2. Awareness campaigns are a necessary strategy to ensure acceptance of, and compliance with, the scheme.
3. A clearly defined monitoring strategy needs to be in place, and the participation conditions must be transparent and adhered to.
4. Given the experience elsewhere in China, it is of the utmost importance to exploit fully one of the most attractive characteristics of PES schemes, that is, their potential financial sustainability. In other words, local authorities must ensure that the funds for agricultural extension services, promotion of organic farming, and compensation to farmers for yields lost to birds, are collected in addition to the current revenues, otherwise the long-term viability of the scheme will be compromised.

References

EEA (1999), *Environmental indicators: Typology and Overview*, pp. 25, European Environment Agency, Copenhagen.

Freeman, A.M. III (1993), *The Measurement of Environmental and Resource Values: Theory and Methods*, Resources for the Future, Washington.

IIED (2006), *Forest Project Summary: Water Chronicles from Costa Rica*, International Institute for Environment and Development.

Landell-Mills, N., and I.T. Porras (2002), *Silver Bullet or Fools' Gold: A Global Review of Markets for Forest Environmental Services and Their Impact on the Poor*, International Institute for Environment and Development, London.

Liu, Z. (2005), 'The Retrospect and Prospects of China's Soil and Water Conservation and Integrated Small Watershed Management', *China Water Resources*, Vol. 19, pp. 17–20.

Markandya, A., P. Harou, L. Bellú, and V. Cistulli (2002), *Environmental Economics for Sustainable Growth*, Edward Elgar, Cheltenham.

Pagiola, S., A. Arcenas, and G. Platais (2005), 'Can Payments for Environmental Services Help Reduce Poverty? An Exploration of the Issues and the Evidence to Date from Latin America', *World Development*, Vol. 33, p. 237.

Pagiola, S., and G. Platais (2003), 'Payments for Environmental Services', *Environment Strategy Notes*, Vol. 3.

Quan R.-C., X. Wen, and X. Yang (2002), Effect of Human Activities on Migratory Waterbirds at Lashihai Lake, China, *Biological Conservation*, Vol. 108, pp. 273–9.

Scherr, S.J., M.T. Bennett, K. Canby, and M. Loughney (2006), 'Lessons Learned from International Experience with Payments for Ecosystem Services: Issues for future PES development in China (draft)', Taskforce on Ecocompensation, China Council for International Cooperation on Environment and Development (CCICED), Beijing.

Scherr, S.J., and J. Smith, (2000), 'Capturing the Value of Forest Carbon for Local Livelihoods: Opportunities Under the Clean Development Mechanism', Notes from CIFOR Bellagio meeting, Unpublished work.

White, A., and A. Martin, (2002), 'Who Owns the World's Forests? Forest Tenure and Public Forest in Transition', in Sayer (ed.), *Earthscan Reader in Forestry and Development*, Earthscan, UK.

Xu, J., R. Tao, Z. Zhiqiang, and M.T. Bennett (2006), 'China's Sloping Land Conservation Programme: Does Expansion Equal Success?', Working paper.

Zhou, J. (2005), 'Holding Back the Desert: Entrepreneurs Have Embarked On a Mission to Protect the Ecology of Arlarshan Plateu', *Beijing Review*, Vol. 48, No. 24.

7

Encouraging Re-vegetation in Australia with a Groundwater Recharge Credit Scheme

WENDY PROCTOR, JEFFERY D. CONNOR, JOHN WARD, AND DARLA HATTON MACDONALD

INTRODUCTION

Widespread clearing of deep rooted perennial native vegetation on individual landholdings in Australia, primarily for agriculture, has occurred over the last 200 years. Environmental consequences have manifested as increased dryland and irrigation related salinity, reduced habitat for native species, rising watertables, and declining water quality in rivers and streams. Past policy approaches to address the adverse environmental consequences of native vegetation clearance have often not motivated land management changes at a scale sufficient to meet mitigation targets. Using a case study approach, this chapter describes a developed methodology to assist in the implementation of a PES framework to combat the effects of native land clearing. This PES approach—a groundwater recharge credit trading scheme—was implemented in the Bet Bet Catchment dryland farming community of north central Victoria, Australia (Map 7.1).

In Australia, the approaches falling under the PES framework are referred to as Market-Based Initiatives (MBIs). MBIs involve regulations or laws that encourage behavioural change through the price signals of markets, as opposed to the explicit directives for environmental management associated with regulatory and centralized planning measures (Stavins 2003). The primary motivation for MBI approaches is that if environmentally appropriate behaviour can be made more rewarding to land managers, then private choice will better correspond to the best social, economic, and

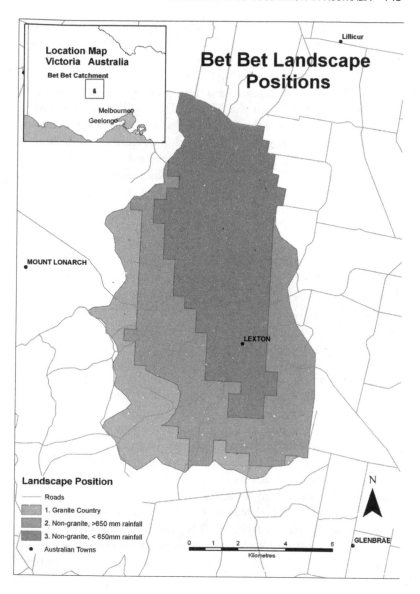

Map 7.1: Map of Bet Bet catchment

environmental outcomes. To encourage development of market-based approaches to water quality and salinity from diffuse sources, the Australian Commonwealth Government allocated funds to 11 MBI pilot projects in 2003 (NAP 2003; Grafton 2005).

The Bet Bet catchment is a relatively small catchment of approximately 9600 ha in the Murray Darling Basin, identified as the major source of more than 40,000 tonnes of salt annually entering the Boort irrigation area from the Loddon dryland catchment areas (Connor et al. 2004). The Bet Bet catchment (which lies in the south west corner of the Loddon River Catchment) was chosen as an area to field test a recharge cap and trade policy because recharge in the area contributes more salt per volume of drainage to local rivers than any other sub-catchment in the region (Clifton 2004).

Figure 7.1 is a schematic representation of rising groundwater recharge levels resulting from land management effects in catchments similar to the Bet Bet. Groundwater recharge increases as an inverse function of the level of deep rooted perennial vegetation (illustrated in panel B). Increased hydraulic pressure in the mound above the saline aquifer causes a subsequent rise in both the water table and the level of salt intrusion in the river system. In the Bet Bet region, the majority of salinity impacts are exported to downstream river districts, where the costs of salinization are incurred primarily by downstream irrigators. Increased volumes of recharge resulting from native vegetation clearance, lead to episodes of increasingly mobilized salt loads in the landscape. The additional salt is exported into connected river systems presenting a risk for the long-term viability of downstream irrigated horticultural and agricultural crops through soil salinization that leads to yield loss. In addition, increased river water salinity levels lead to accelerated infrastructure degradation (Clifton 2004), and threaten the functional organization of downstream riparian ecosystems (Overton and Jolly 2004).

Recharge rates and associated rates of salt mobilization in the area depend on the regional geomorphology with localized fractured rock conducive to high groundwater recharge rates. In addition, the rate of recharge and thus external salinity impacts, depend on the type of vegetation ground cover and farm-specific cropping, grazing, and management decisions. Extensive replacement of deep rooted woody perennials and perennial pasture with shallow rooted annual pastures has been identified as a key factor in increased rainwater soil percolation and subsequent groundwater recharge.

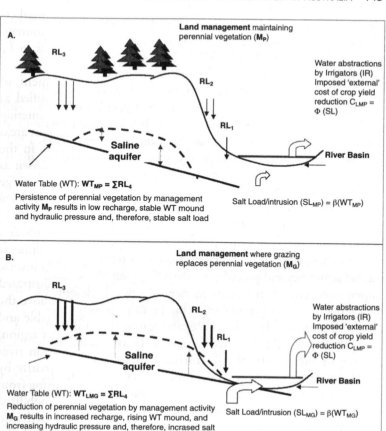

Fig. 7.1: The salinity problem—schematic of the hydro-geology of irrigation water quality affected by variable upper catchment salt loads

The dryland salinity problems explained above are becoming increasingly common across agricultural regions in Australia. This pilot project was designed to test an MBI approach to motivate re-vegetation efforts and thus reduce consequent groundwater recharge. The objective of the trial was to develop and test the feasibility of a recharge credit scheme to provide flexible incentives to motivate more cost-effective re-vegetation efforts, to reduce consequent groundwater recharge, mobilized salt loads, and eventual levels of river salinity.

Tradeable permit schemes for managing environmental problems are becoming more widely accepted by policy makers in Australia,

North America, and elsewhere (Randall 2003; Sterner 2003; Harrington et al. 2004). Subject to controversy and debate ten years ago (Keohane et al. 1998), MBIs have evolved to the point of becoming received wisdom in many environmental policy circles (Stavins 2003). The National Action Plan for Salinity and Water Quality and the National Heritage Trust exemplify a Commonwealth impetus for the increasing application of market-based solutions in Australia. Despite this increasing acceptance, Tietenberg (1998, 1999) concludes that many tradeable permit schemes have failed because of inadequate attention to ex ante instruments and institutional design. To date, a priori prescriptions of alternate market institutions and auction systems, calibrated to catchment specifications, enabling the reliable translation of market theory to an operational reality, have not yet emerged. The outcome for managing authorities may be the hasty adoption and implementation of potentially inappropriate market structures and procedures, often to expedite and satisfy policy imperatives. Any adverse consequences of a poorly designed scheme may remain undetected for long time periods, possibly eroding the potential economic benefits and exacerbating the problem that the change was originally intended to resolve. Inappropriate design may also negate the opportunity for further innovation.

This chapter describes a novel methodology used in the design and evaluation of the Bet Bet recharge trading scheme. The approach involved ex ante identification and evaluation of a complete range of potential impediments to the effective functioning of a market for the exchange of tradeable recharge credits. Experimental economics settings were framed by a synthesis of salient biophysical, economic, and attitudinal characteristics and prevailing social norms of the catchment (Ward et al. 2007). Experimental treatments measured and evaluated behavioural responses to alternative cap and trade solutions and voluntary, community-crafted compacts for recharge management. The ex ante design and testing methodology used in this trial represents an emerging systematic process for policy makers to gain confidence, experience, and expertise in the design and testing of a cap and trade policy prior to its implementation. Until recently, in-depth appraisals of the potential inclusion and capacity of cap and trade instruments in Australian policy portfolios have been limited. This chapter demonstrates how the design and testing methodology empirically informed on-ground policy implementation, including detailed specification of landholder obligations to manage recharge, credit accounting and trading rules, monitoring protocols and non-compliance enforcement.

ACTORS AND ECOSYSTEM SERVICES INVOLVED

The trial is designed to demonstrate that a market-based approach to achieving land use change is a viable alternative to the current government system of regulatory approaches, and input-based payment incentives. In the past, government-sponsored efforts to motivate changes in management regimes on privately tenured land have relied on traditional farm extension processes, legal and statutory remedies and a scheme providing scheduled payments for the re-establishment and management of deep rooted perennials. Despite regional promotion, the level of established re-vegetation and consequent groundwater recharge and river salinity, have not been at a scale sufficient to comply with prescribed salinity targets (Connor et al. 2004).

The status quo property rights arrangement in Victoria where the trial is being implemented involves no explicit requirements for dryland farmers to meet water quality requirements or to manage levels of recharge resulting from their practices. Nor are there currently any well-defined and enforceable arrangements that would allow those who may suffer adverse consequences of increased salinity to compensate farmers causing salinity to reduce impacts. In essence what has existed is an implicit but poorly defined right of dryland farmers to manage recharge as they like. The trial scheme was, therefore, implemented in an attempt to achieve better outcomes than previously administered instruments, constrained by extant property right regimes. The trial is expected to run for two years.

The Bet Bet trial relies on voluntary participation in a process designed to demonstrate how altered individual land use decisions can contribute to collective outcomes that reduce the aggregate impact of salinity. The Bet Bet community comprises approximately 130 landholders, 17 of whom have agreed to enter into individual contracts to change land management actions on their properties and comply with individually specified and contractually obligated recharge targets. Collectively, these individuals contribute to a catchment-wide (community) goal for aggregate recharge reduction.

Overall, the market elements of this trial, chosen through the trial design process, include:

1. community agreement to achieve a specified level of recharge control;
2. individual landholder contracts to achieve a specified level of recharge control in return for payment;
3. trading of excess recharge credits between landholders in order for all landholders to meet their contract obligations;
4. bonus payments to landholders who exceed their recharge control targets; and
5. a community bonus if the catchment target is met or exceeded.

A tradeable recharge right involves establishing an enforceable, prescribed threshold of aggregate recharge attributable to the Bet Bet catchment, distributing entitlements amongst recharge sources as a specified number of units and allowing trade of those units among scheme participants. To satisfy compliance obligations, each participant in the scheme must be able to surrender units equal to their entitlement at the end of an accounting period of two years. Therefore, participants can choose to alter land actions in response to individual management capacity, landscape attributes, and production costs. Alternatively, those in deficit can secure additional recharge units from those in surplus through market exchange. Compliance is, therefore, defined in terms of a resource use cap rather than direct requirements for delivery of a service. The link between the two, however, is explicitly calculated and recorded when performance assessments are carried out.

Performance is assessed by monitoring vegetative groundcover at the end of each cropping year in December. The functional relationship between vegetation type, management, landscape position, and groundwater recharge in the Bet Bet catchment has been previously established as part of the instrument design process (Connor et al. 2004; Clifton 2004). Audits, measuring the percentage of groundcover that landholders actually achieved, are conducted and the results used to compute an empirically based estimate of the recharge volume for each landholder. A credit surplus or deficit position is assigned based on the audited cover that each landholder has achieved relative to the level of credits that they committed to provide.

An independent 'auditor' is involved in the trial to audit pasture groundcover and tree establishment performance. For pastures, the auditor takes multiple measurements to provide a representative sample for each paddock. Measurements are taken using a 500 x 500 mm square with four evenly spaced, horizontal and vertical strings or wires. The quadrant wires intersect at 16 points within the square (Figure 7.2). Cover is assessed by placing the square on the pasture, grass, or crop and then counting the number of intersections that lie directly above the green vegetation. The percentage of those 16 intersections that sit above green vegetation is the cover at that point.

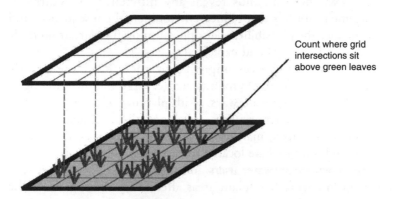

Count where grid intersections sit above green leaves

Fig. 7.2:   Measurement of cover for pastures, crops, and understoreys

An additional actor involved in the trial is an independent 'broker' who is engaged to maintain the recharge accounts for each landholder. These accounts show:
1. the obligation agreed to;
2. the audited estimated recharge;
3. the reference level of recharge; and
4. the number of credits in excess or in deficit of the reference level.

The broker is also able to facilitate the creation and trading of credits. Credits can be created by undertaking additional perennial plantings within the target catchment. Trading of credits facilitated by the broker can also take place for landholders holding salinity recharge credit surpluses or deficits. Salinity recharge credits can be traded at any price negotiated by the landholders.

While capping recharge imposes a cost on individuals, the opportunity to trade has the potential to compensate that loss or reduce the cost burden. Some individuals will choose to use more than their quantum (and incur a debit), and others will choose to use less (being rewarded with credits). The brokerage feature is an approach to overcome the policy challenge to create the opportunity for a 'frictionless' market setting where participants can quickly learn to understand the advantages of trade with low learning and exchange costs relative to trade benefits. To the extent that brokerage reduces market friction, savings to landholders through market exchange between individuals with surplus credits and those in deficit may be considerable. Brokerage can increase the level of information from market exchange and thus reveal any differences in returns to management options that reduce environmental consequences and thus enhance the probability that opportunities for gains from trade are quickly discovered and exploited.

The ecosystem services improved by the scheme are directly related to reduced levels of groundwater recharge and the lowering of salt levels in soils and waterways. With planting of deep rooted tree species, other ecosystem services, such as biodiversity and provision of shade for livestock, may also be improved. Spatially, the services affected and improved are located in the Bet Bet catchment but also include downstream water users and those who enjoy the amenity values of low lying floodplain areas affected by recharge from the Bet Bet region. Participants in the scheme, however, are only those who reside in the Bet Bet catchment and volunteer to take part under the conditions of the contract. The eventual beneficiaries may be downstream and not necessarily reside in the catchment.

THE IMPLEMENTATION PROCESS

Implementing the scheme takes place based on the contract agreed to by the landholder to establish and maintain perennial plantings in ways capable of reducing recharge in the landscape. In essence, landholders will receive payment in exchange for their actions to change land-use. The actual agreements related to land-use change are restricted to the types of plantings involved. Five possibilities exist including: (i) low density farm forestry; (ii) high density farm forestry; (iii) native tree establishment; (iv) phalaris perennial pasture, or (v) lucerne perennial pasture. Payment is based on performance and includes an establishment (initial) payment, and a management (subsequent) payment(s) based on monitored performance in the following years. Payment is on recharge credits calculated on a per hectare basis. Table 7.1 describes the levels of credits per hectare that can be achieved given monitored performance levels for each specific practice. The actual level of assigned credits depends on the level of cover of the pastures and/or the number

Table 7.1:   Recharge credits according to audited ground cover landscape position and annual rainfall

Practice (audited performance /ha)	Zone 1 (700mm+)*	Zone 2 (650mm+)	Zone 3 (600mm+)
Low density Farm Forestry (200 stems)	6.1	4.9	3.7
High density Farm Forestry (600 stems)	9.6	7.6	5.7
Low density Farm Forestry (180 stems)	5.4	4.3	3.2
High density Farm Forestry (540 stems)	8.4	6.8	5
Low density Farm Forestry (160 stems)	4.5	3.5	2.6
High density Farm Forestry (480 stems)	7	5.5	4.1
Low density Farm Forestry (140 stems)	3.5	2.7	2
High density Farm Forestry (420 stems)	5.5	4.3	3.2
Native Tree Establishment (600 stems)	21	16.9	12.7
Native Tree Establishment (540 stems)	19	15.1	11.3
Native Tree Establishment (480 stems)	15	12.1	9.1
Native Tree Establishment (420 stems)	12	9.4	7.1
Phalaris Pasture (100 per cent cover)	2.4	2.4	2.4
Lucerne Pasture (100 per cent cover)	4.4	4.2	3.5
Phalaris Pasture (90 per cent cover)	2	2	2
Lucerne Pasture (90 per cent cover)	3.5	3.5	3.5
Phalaris Pasture (80 per cent cover)	1.6	1.6	1.6
Lucerne Pasture (80 per cent cover)	3	3	3
Phalaris Pasture (70 per cent cover)	1.2	1.2	1.2
Lucerne Pasture (70 per cent cover)	2.5	2.5	2.5

Notes: *annual rainfall
Source: Clifton (2004).

of stems on tree plantings measured each year as described in the previous section.

Based on the ex ante analysis, approximately 3000 credits are sought from tree-based (forestry, native tree establishment) practices. Provided 90 per cent (2700 credits) of the tree-based target is met, up to 1000 credits will be available for perennial pasture (phalaris/lucerne) establishment. In total, $38.50 is offered to landholders for each unit of recharge they control over the life of the project. If a total of 3750 recharge units are produced then there will be a communal 'bonus' payment of $7500.

Table 7.2 gives an example of what outcomes of participation in such a tradeable credit recharge scheme could look like for four different landholders in the Bet Bet region.

At the outset there were concerns that low returns to some farmers that had resulted in near zero enrolment in the existing scheduled

Table 7.2: Schematic representation of salinity recharge credit trial functioning for an illustrative example

		Landholder 1	Landholder 2	Landholder 3	Landholder 4
		contracts to convert pasture to farm forestry	contracts to convert degraded annual pasture to perenial lucern	contracts to convert pasture to native vegetation	contracts to convert degraded annual pasture to perenial lucern
Contract Negotiation	Baseline recharge under current landuse, Qi^b =	120 ML	200 ML	150 ML	300 ML
	Expected recharge for successfully established farm forestry =	15 ML	70 ML	10 ML	100 ML
	Obligation under contract, Qi^0 =	105 ML	130 ML	140 ML	200 ML
	Start-up incentive = $50/ML*Qi^0$ =	$5,250	$6,500	$7,000	$10,000

Year of implementation: the season turns out to be low rainfall

Audit and Credit/ Debit Accounting	Landholder 1's audit = 75 per cent of stems required to fulfil obligation	audited reduction = 78.5ML			
	Landholder 2's audit = 80 per cent of cover required to fulfil obligation		audited reduction = 104ML		
	Landholder 3's audit = 60 per cent of stems required to fulfil obligation			audited reduction = 84ML	

(contd...)

Table 7.2 (contd...)

		Landholder 1	Landholder 2	Landholder 3	Landholder 4
		contracts to convert pasture to farm forestry	contracts to convert degraded annual pasture to perenial lucern	contracts to convert pasture to native vegetation	contracts to convert degraded annual pasture to perenial lucern
	Landholder 4's audit = 65 per cent of cover required to fulfil obligation All obligation seasonally adjusted by weighted avg performance of 69 per cent of obligation				audited reduction = 130ML
	debit/credit = audited seasonally adjusted recharge— obligation	+6 ML	+14ML	−12ML	−8ML
Trading	Landholder 1 sells 6 credits to Landholder 4	0 ML balance	+ 14 ML balance	−12 ML balance	−2 ML balance
	Landholder 2 sells 12 credits to Landholder 3	0 ML balance	+ 2 ML balance	0 ML balance	−2 ML balance
	Landholder 2 sells 2 credits to Landholder 4	0 ML balance	0 ML balance	0 ML balance	0 ML balance
	Bonus paid equal to 20 per cent of start-up in proportion to contribution to total recharge	$1,140	$1,510	$1,220	$1,885

Source: Authors.

payment scheme prior to the cap and trade trial would also lead to limited enrolment in the cap and trade scheme. In fact, enrolment increased from only 5 ha in 2004 using a standard input-based payment to over 100 ha in the 2005 cap and trade scheme. The relatively small scale of the programme, viz. 17 participants, is a result of an intentionally limited number of targeted potential participants and limited budget for this trial programme.

ADDITIONAL INCENTIVES INVOLVED

The voluntary nature of the credit trading scheme, contingent on the lack of articulated property right obligations for recharge management, means that non-compliance within the Bet Bet catchment does not attract any sanctions if people do not wish to participate. Vatn and Bromley (1995), Ostrom (1998), and Gintis (2000) argue that non-monetary rewards and motivations such as prestige, public recognition, group belonging, avoidance of group sanction, and desire to contribute to the public good can all represent powerful motivators in some contexts. There has been considerable theoretical work suggesting that policies involving collective outcome-based payments or penalties can motivate high rates of environmental action and cost effectiveness in certain settings (Segerson 1988; Isik and Sohngen 2003; Ipe et al. 2001). In particular, previous research (for example, Ostrom 1998; Gintis 2000; Tisdell et al. 2004) reports willingness to diverge from individualistic profit maximizing behaviour for the public good in small, cohesive communities. Poe et al. (2004) posit that a free riding problem can arise with collective incentive policy where there is too little individual incentive and individual behaviour is not easily observed.

Given the small cohesive nature of the Bet Bet community revealed in social survey results (Connor et al. 2004), a policy designed to harness the potential power of pro-social motivations in the trial area may have potential to increase trial enrolment. Experimental economics results (Ward et al. 2007) suggest that a collective payment could have the potential to address the risks of low enrolment given relatively flat pay-offs to recharge reducing practices and informational challenges associated with understanding pay-offs. Thus, a feature of the scheme is a community level payment in addition to individual payments, compensating establishment and opportunity costs. The community incentive is paid in the form of a community bonus if an aggregate recharge reduction target is met or exceeded. This type of scheme attempts to harness peer group pressure to ensure that each individual meets the contracted target so that the entire community benefits. There is also additional community-based motivation for highly targeted, non-bidding members of the Bet Bet farming community to take part in the scheme.

PERMANENCE, MONITORING, AND ACCOUNTING

Mechanisms have been put in place to ensure that the effects and benefits of the trial will extend into the future. One mechanism to favour more permanent action assigns credit levels and, therefore, payments based on two factors:

- Expected permanence—practices expected to be more permanent are given more credit (native vegetation protection is assumed to be more permanent than farm forestry which is assumed to be more permanent than perennial pasture).
- Expected annual recharge reduction—the expected recharge reduction for each practice has been estimated using a crop water balance model and are calculated relative to a defined baseline level of recharge equal to the estimated recharge under perennial pasture with 70 per cent December cover.

With regards to permanence however, a lack of well-defined property rights represents a significant impediment. Without a more explicit definition of either farmer obligations to manage groundwater recharge or rights to contract water quality improvement for those adversely impacted by dryland recharge induced salinity, no formal and permanent market for recharge credit can be established.

A feature of the scheme to encourage more persistent action is a contract including not only commitments to establish perennial vegetation but also commitments to maintain plantings in a manner that will provide permanent recharge outcomes. While the former programme in the area was based on input payments with no incentive for ongoing management to achieve environmental goals, the cap and trade policy rewards landholders who persist with management with performance-based incentive payments.

Shortle and Horan (2001) and Schary (2003) argue that developing policies capable of realizing savings by focusing on performance coupled with compliance flexibility is challenging for diffuse source pollution because monitoring actual outcomes is often technically infeasible or very costly. This represents a substantial challenge to effective cap and trade schemes to address diffuse source environmental issues such as salinity. To effectively participate in the exchange of tradeable recharge credits, land managers need accounting and auditing that allows an evaluation of their management decisions prior to implementation and monitoring of progress against their targets or commitments. Similarly, administrators of the scheme must also have the capacity to monitor and audit the outcomes of changes in land use or management practices and to attribute change in recharge to either landholder action or climate. Since groundwater recharge and salinity are not readily measured directly, a prerequisite to implementing a cap and trade is the development of a reliable and transparent surrogate indicator to assist all participants in evaluating recharge and salinity impacts of land management actions.

Thus, a first step in this project was the development of robust and community validated biophysical and hydrological modelling to provide information about groundwater recharge rates as a function of variable vegetation cover and land management at the farm scale, differentiated according to landscape position (Clifton 2004; Connor et al. 2004).

The resulting crop water model accounts for:
1. differences in rates of annual and perennial crop and native tree evapo-transpiration;
2. temporal differences between tree and crop types for maximum transpiration rates to be realized; and
3. differences in recharge reduction resulting from landscape position (differential recharge reduction is a function of rainfall, slope, soil permeability, and levels of fractured granite and soil transmissivity).

In the model (illustrated in Map 7.1) R_i represents the recharge rate for farm i, managing crop j, where:

$$R_i = (C_{ij}, A_{ij}, R_{Ai}, G_i, L_k): \text{ and} \qquad \ldots (7.1)$$

C_{ij} is crop type and management
A_{ij} is area of crop type
R_{Ai} is annual rainfall
G_i is soil type and geomorphology
L_k is landscape position, k=1, 2 or 3 where:
        k=1 represents lower slope;
        k=2 represents break of slope;
        k=3 represents ridge and upper slope.

$j = 1–5$, where:

$j=1$ represents annual grazing;

$j=2$ represents perennial pasture (phalaris set grazing);

$j=3$ represents perennial pasture (phalaris rotational grazing);

$j=4$ represents native tree vegetation;

$j=5$ represents farm forestry (less than 10 years old).

C_{ij}, A_{ij} represent endogenous variables in a farm decision set; and R_{Ai}, G_i, L_k represent exogenous variables in a farm decision set.

The model developed accounts for three key biophysical determinants of recharge differences across locations and actions shown in Map 7.1.

1. Ceteris paribus, for crop j, recharge from lower slope (L_1) is less than the recharge from break of slope (L_2) which is less than recharge from upper slopes (L_3), viz. $RL_1 < RL_2 < RL_3$.

2. Increased deep rooted perennial vegetation reduces groundwater recharge: viz. for landscape position L_k, subject to land management regime M_P (Panel A) or M_G (Panel B), recharge R_i is such that $R_i\, L_k\, M_P < R_i\, L_k.\, M_G$.

3. The estimated costs of groundwater recharge for land management activity at farm i, at landscape position L_k is such that:

   $R_{MG} > R_{MP}$; recharge from annual grazing is greater than recharge from perennial grazing or forestry;

   $WT_{MG} > WT_{MP}$; the water table level is higher for annual grazing than perennials;

   $SL_{MG} > SL_{MP}$; salt load is greater for annual grazing than perennials;

   $C_{MG} > C_{MP}$: incurred irrigation costs are greater for annual grazing than perennials.

Another impediment to establishing a robust, permanent, recharge exchange scheme is the potential for thin markets. One of the conditions necessary for efficient, functioning, competitive markets is a sufficient number of traders to ensure that no one participant can influence the terms on which transactions occur. Thin markets are characterized by small numbers of buyers and sellers. A limited number of buyers and sellers introduces the potential for credit trade market failure in a number of ways including price volatility and restricted supply (Stavins 1995; Kampas and White 2003), spatial concentration of permits, permit hoarding and a potential impedance of new market entrants (Tietenberg 1998), a lower probability of satisfying market needs associated with increased transaction costs (Stavins 1995), and unreliable recharge outcomes (Dinar and Howitt 1997). Goodstein (2002, p. 330) argues that the United States EPA's emissions trading programme, introduced in 1976, floundered because of thin markets and concerns about permit hoarding and spatial concentration of effluents. Stavins (1995) argues that a thin market reduces market efficiency by raising the relative costs of transactions: fewer participants implies a lower probability, both real and perceived, of finding trading partners to resolve market demands, while transaction costs remain constant or increase. To date, the trial focus has been on a relatively small area with relatively few participants. Smith (1982) argues a countervailing view, noting experimental economics findings that suggest that the numbers in the trial are sufficient to avoid thin market problems.

In the design phase it was recognized that the problem of thin markets could be exacerbated by the impact of random variation in weather conditions on the success of actions to reduce recharge. For example, many years of drought, would tend to lead to higher rates of establishment failure. Given that effects of weather on credit surpluses or deficits would tend to be correlated across years for participants within the geographically small trial area, potential for excess credit supply or demand within seasons was seen as a factor that could lead to people deciding not to participate in future market schemes. To overcome this impediment, banking and borrowing of credits is allowed. Goodstein (2002) found that in the US unleaded gasoline refinement quota cap and trade system, temporal flexibility implemented through credit banking was a key reason for the programme's cost effectiveness. Thus in the Bet Bet trial, a participant who has credits in excess of obligations after annual performance monitoring and salinity account reconciliation, can bank or set aside credits to offset debits in future years. Any participant with credits banked in previous years can use them against debits to balance a current salinity account.

POTENTIAL PROBLEMS

In ideal markets where there are no transactions costs, trade takes place whenever there is potential for marginal gains and profit. Real markets for tradeable emissions credits, including recharge, require substantial investment of time by participants to understand potential gains, seek out trading partners, and negotiate trades (Randall 2003). A potential impediment to credit trade results if potential gains from trade are low compared to transaction costs. Newell and Stavins (2003) and Sterner (2003) have identified small potential gains from trade arising from relatively small differences in marginal abatement costs across sources as an impediment to credit trade policy in point source contexts. Vanclay (2004) and Barr (1999) argue that financial returns are only one factor in the utility function determining farm management choices. When there is relatively little difference in pay-offs across management practice, non-financial considerations such as family lifestyle are often key determinants of management practice choice.

The potential gains from trade in the salinity credit trade trial and returns arising from recharge reduction options across land management practices and landscapes within the Bet Bet were modelled. Results indicated that whilst there is sufficient differentiation in marginal abatement costs across practices and location, potential gains from trade are less than 10 per cent of total revenues (Connor et al. 2004). Field interviews (Connor et al. 2004) suggest that some practices which farm economics modelling indicates can reduce recharge and improve returns, may be less attractive than purely economic considerations would indicate. More careful rotational grazing and other perennial pasture management in particular are effective in reducing recharge and have the potential for slightly greater per hectare returns, but require much more management effort.

Ultimately, potential for gains to credit trade are limited by differences in physical productivity and opportunity costs among potential trade participants. The amount that is actually realized can be influenced by policy design. Ensuring that participants realize potential gains from trade involves designing an administratively efficient method of monitoring and recharge accounting. Policy initiatives to ensure the introduction of cost effective monitoring and accounting schemes that are transparent, consistent, and credible to all participants predicate successful cap and trade schemes. Accounting conventions establish a clear link between land management actions at appropriate scales and the consequent environmental outcome. As the functional relationships between river salinity, groundwater recharge, and land actions are not readily visible, a proxy indicator inclusive of re-vegetation type, success of establishment, and maintenance was imputed in this case study.

When the differential in marginal abatement costs is small, expressed as a flat pay-off function, Pannell (2006) proposes that the manner of presentation and treatment of information to potential adopters can be a key determinant of the level of practice adoption. Strong social networks and membership in catchment groups (Cary et al. 2002), as well as extension and promotion programmes (Marsh et al. 2000) have been shown to be important determinants of conservation practice uptake in flat pay-off function settings. This suggests that in trial implementation, information and credit trade policies are likely to be complementary. A design solution to the impediment of a flat pay-off function with little information is information provision.

In all such schemes, there are also issues related to concerns of equity and fairness for all members of the catchment community being targeted. For example, there can be considerable equity concerns related to the method of entitlement distribution in cap and trade schemes. Perman et al. (1999, pp. 316–17) note that the initial distribution of property rights determines the division of the net gains which accrue to the negotiating parties. Thus, a major challenge in designing a credit trade system is the mechanism used to allocate the initial permits to individuals (Baumol and Oates, 1988). Auctions and free distribution (grandfathering) are the two main procedures employed by governing agencies in the allocation of transferable resource permits. Revenues from auctioning of permits go to the state, whereas the benefits gained from the grandfathering accrue to those granted the entitlement (Tietenberg 1999).

The status quo ante in the catchment is an upfront payment for implementation of a practice (such as a payment per hectare of trees planted). A primary focus of the trial is to test a performance-based system with payments based on the outcomes of practices (for example, payment on success rate of establishment of tree planting). The shift from the status quo ante to the trial approach will change expected costs, income, and income variability. With the status quo ante, the risk of less than intended recharge reduction (through planting establishment failure) is assigned to the Government. Changing to a performance-based system shifts that risk to farmers. There are two risks relating to returns that participants could face in a performance based system: (i) risk associated with random, exogenous events such as rainfall variability and (ii) the risk of management related failures of options in achieving recharge reduction.

Given that the trial involves voluntary participation and that the status quo ante programme continues to run simultaneously, there is potential for low enrolment rates to impede functioning if the trial involves greater risks to participants than the status quo ante without commensurate improvements in potential returns.

If a fundamental change in property rights that created limits on dryland farm recharge could be implemented, distribution of recharge rights by auction would be possible. As mentioned already, this is the mechanism that economic theory suggests is the most economically efficient. Within the current property rights framework without any explicit limits on recharge, the only possibility is to begin by grandfathering the right to the current level of recharge and offering payment for improved practice or performance. A tension arises because the approach that would create the superior performance incentive, payment on outcome, involves significant risk in comparison to the status quo ante incentive, payment on implementation of practice. Low participation rates are a likely outcome. The compromise solution implemented in the trial, involved a partial payment on establishment and a partial payment based on audited performance outcome.

Conclusions and Next Steps

Rigorous comparison of trial results with the status quo ante scenario to ascertain if significant salinity reduction has occurred is still too early to be carried out. The trial commenced in the first quarter of 2005 and the first performance appraisals of farms occurred around December with audits being completed around the first quarter of 2006 (one year after the first contracts were signed).

Upon completion of the trial in mid-2006, the success of this scheme relative to previous instruments and recharge management policies will be modelled and compared. Additionally, the beneficial effects of tree planting may not start to be realized until several years after the start of the trial.

Although it is still too early to gauge the success or otherwise of the current trial, some policy related issues can be highlighted as a result of the review of the scheme provided in this chapter, guiding and informing future decisions.

One problem is the geographically constrained trial area which can potentially lead to the adverse effects of thin markets. Conceptually, the most obvious approach to overcoming thin market problems would be expansion of the scale of the trial to include more participants. If this involved an expanded geographic area, it is also possible that potential for weather related excess credit supply or demand leading to thin markets would be reduced. Credit banking and borrowing across years as exists in the current trial is one approach to reducing thin market problems arising from weather related excess credit supply or demand. As the trial is only over two years, it will be difficult to assess the effectiveness of credit banking but this ability should be subject to greater scrutiny if the trial is allowed to continue longer.

Theoretically, tradeable recharge entitlements can be assigned to either one of the negotiating parties: farmers (as the source of recharge) or the downstream beneficiaries of recharge reduction. At present, the trial only involves participants from the Bet Bet catchment— the initiators of the salinity problems. The current property rights framework allows farmers to manage recharge without regard to external effects. Participants are induced to cap recharge through an incentive payment. This is in contrast to the more common cap and trade approach, reliant on statutory obligations. A future option may seek to develop specified property rights for clean water for those harmed by salinity or recharge management obligations for those who create salinity. This would involve establishing legally defined and enforced limits on recharge rights or some proxy for recharge such as inputs, outputs, or practices correlated with groundwater emissions as a property rights basis for the tradeable recharge policy trial.

As mentioned previously, some of the benefits and costs of the land-use change induced here may be impacting outside the trial area and so some of the more important participants who should take part in the scheme are actually excluded. Redefining property rights could overcome the impediments of thin markets by engendering wider participation from agents characterized by a greater differential in costs of salinity abatement.

Finally, a thorough comparison of this scheme with the previous incentive scheme, where farmers were rewarded with cash payments for beneficial land-use change, is required. Careful measurement of the biophysical benefits, revenue implications for farmers, and the cost effectiveness from the perspective of the implementing agency is required, including evaluation of the transaction, administration (including brokerage), and monitoring costs.

REFERENCES

Barr, N.F. (1999), *Salinity Control, Water Reform and Structural Adjustment: The Tragowel Plains Irrigation District*, Doctoral thesis, Institute of Land and Food, University of Melbourne, Melbourne, Australia.

Baumol, W.J. and W.E. Oates (1988), *The Theory of Environmental Policy*, 2nd edn, Prentice-Hall, New Jersey.

Cary J., T. Webb, and N. Barr (2002), *Understanding landholders' capacity to Change to Sustainable Practices: Insights About Practice Adoption and Capacity to Change*, Agriculture, Fisheries and Forestry Australia, Canberra, Australia.

Clifton, C. (2004), *Water Salt Balance Model for the Bet Bet Catchment*, SKM, Bendigo, Victoria.

Connor, J., J. Ward, D. Thomson, and C. Clifton (2004), *Design and Implementation of a Landholder Agreement for Recharge Credit Trade in the Upper Bet Bet Creek Catchment*, Victoria. CSIRO Land and Water Report S/04/1845.

Dinar, A. and R.E. Howitt (1997), 'Mechanisms for Allocation of Environmental Control Cost: Empirical Tests of Acceptability and Stability', *Journal of Environmental Management*, Vol. 49, No. 2, pp. 183–203.

Gintis, H. (2000), 'Beyond Homo Economicus: Evidence from Experimental Economics', *Ecological Economics*, Vol. 35, No. 3, pp. 311–22.

Goodstein, E.S. (2002), *Economics and the Environment*, 3rd edn, John Wiley and Sons Inc., New York.

Grafton, Q. (2005), *Evaluation of Round One of the Market-based Instrument Policy Trial*. Asia Pacific School of Economics and Government. Australian National University, Canberra, ACT.

Harrington, W., R.D. Morgenstern, and T. Sterner (2004), *Choosing Environmental Policy: Comparing Instruments and Outcomes In the United States and Europe*. Resources For The Future. Washington, DC.

Ipe, V.C., E.A. DeVuyst, J.B. Braden, and D.C. White (2001), 'Simulation of a Group Incentive Programme for Farmer Adoption of Best Management Practices', *Agricultural and Resource Economics Review*, Vol. 30, No. 2, pp. 139–50.

Isik, H.B. and B. Sohngen (2003), *Performance-based Voluntary Group Contracts for Nonpoint Source Pollution*, Proceedings of the American

Agricultural Economics Association Annual Meeting, 27–30 July 2003, Montreal, Canada.

Kampas, A., B. White (2003), 'Selecting Permit Allocation Rules for Agricultural Pollution Control: A Bargaining Solution', *Ecological Economics*, Vol. 47, Nos. 2 and 3, pp. 135–47.

Keohane, N., R. Revesz, and R. Stavins (1998), 'The Choice of Regulatory Instruments in Environmental Policy,' *Harvard Environmental Law Review*, Vol. 22, No. 2, pp. 313–67.

Marsh, S., D. Pannell, and R. Linder (2000), 'The Impact of Agricultural Extension on Adoption and Diffusion of Lupins as a New Crop in Western Australia', *Australian Journal of Experimental Agriculture*, Vol. 40, No. 4, pp. 571–83.

MBI Working Group on Market-based Instruments (2002), *Investigating New Approaches: A review of Natural Resource Management Pilots and Programmes in Australia that Use Market-based Instruments*, National Action Plan for Salinity and Water Quality, Canberra.

Newell, R.G. and R.N. Stavins (2003), 'Cost Heterogeneity and the Potential Savings from Market-Based Policies', *Journal of Regulatory Economics*, Vol. 23, No. 1, pp. 43–59.

Ostrom, E. (1998), 'A Behavioural Approach to the Rational Choice Theory of Collective Action', *American Political Science Review*, Vol. 92, No. 1, pp. 1–22.

Overton, I. and I. Jolly (2004), *Integrated Studies of Floodplain Vegetation Health, Saline Groundwater and Flooding on the Chowilla Floodplain, South Australia*, CSIRO Land and Water Technical Report No. 20/04.

Pannell, D.J. (2006). 'Flat-earth Economics: The Far-reaching Consequences of Flat play-off Functions in Economic Decision Making', *Review of Agricultural Economics*, Vol. 28, No. 4, pp. 553–66.

Perman, R., Y. Ma, J. McGilvray, and M. Common (1999), *Natural Resource and Environmental Economics*, 2nd edn, Essex, UK, Pearson Education Limited.

Poe, G., W. Segerson, S. Jordan, and C. Vossler (2004), 'Exploring the Performance of Ambient Based Policy Instruments when Non-point Source Polluters Can Cooperate', *American Journal of Agricultural Economics*, Vol. 86, No. 5, pp. 1203–10.

Randall, A. (2003) 'Market-Based Instruments—International Patterns of Adoption, Remaining Challenges, and Emerging Approaches', Proceedings of the 6th Australian Agricultural and Research Economics Society (AARES) Annual National Symposium, 3 September 2003, Rural Industries Research and Development Corporation, Canberra.

Schary, C. (2003), 'Can the Acid Rain Programme's Cap and Trade Approach Be Applied to Water Quality Trading?', Proceedings from the Market Mechanisms and Incentives: Applications to Environmental

Policy Workshop, 1–2 May 2003, United States Environmental Protection Agency, Washington, DC.

Segerson, K. (1988), 'Uncertainty and Incentives for Non-Point Source Pollution', *Journal of Environmental Economics and Management*, Vol. 15, No. 1, pp. 87–98.

Shortle, J.S. and R.D. Horan (2001), 'The Economics of Non-point Pollution Control', *Journal of Economic Surveys*, Vol. 15, No. 3, pp. 255–89.

Smith, V.L. (1982), 'Markets as Economizers of Information: Experimental Examination of the Hayek Hypothesis', *Economic Inquiry*, Vol. 20, No. 2, pp. 165–79.

Stavins, R. (1995), 'Transaction Costs and Tradeable Permits', *Journal of Environmental Economics and Management*, Vol. 29, No. 2, pp. 133–48.

—— (2003) '*Market-based Instrument Policies: What Can We Learn from U.S. Experience (and related research)*', Presented at Twenty years of Market-based Instruments for Environmental Protection: Has the promise been realised?, 23–24 August 2003, Donald Bren School of Environmental Science and Management, University of California, Santa Barbara.

Sterner, T. (2003), *Policy Instruments for Environmental and Natural Resource Management. Resources for the Future*, World Bank and the Swedish International Development Cooperation Agency (Sida) Washington DC.

Tisdell, J., J. Ward, and T. Capon (2004), Impact of Communication and Information on a Complex Heterogeneous Closed Water Catchment Environment, *Water Resources Research*, Vol. 40, No. 9, pp. 1–8.

Tietenberg, T. (1998) 'Ethical Influences on the Evolution of the US Tradable Permit Approach to Air Pollution Control', *Ecological Economics*, Vol. 24, No. 2, pp. 241–57.

Tietenberg, T.H. (1999), 'Tradable Permit Approaches to Pollution Control: Faustian Bargain or Paradise Regauned?', in M.D. Kaplowitz (ed.), Property Rights, Economics, and the Environment, JAI Press, US.

Vanclay, F. (2004), 'Social Principles for Agricultural Extension to Assist in the Promotion of Natural Resource Management', *Australian Journal of Experimental Agriculture*, Vol. 44, pp. 213–22.

Vatn, A. and D. Bromley (1995), 'Choices Without Prices Without Apologies', in D. Bromley, (ed.), *The Handbook of Environmental Economics*, Cambridge, Massachusetts: Blackwell Publishers, pp. 3–25.

Ward, J., J. Connor, and J. Tisdell (forthcoming), 'Aligning Policy and Real World Settings: An Experimental Economics Approach to Designing and Testing a Cap and Trade Salinity Credit Policy', in J. Shogren, and T. Cherry (eds), *Experimental Methods in Environmental Economics*, Oxford: Routledge.

8

Marketing Ecosystem Services

MANUELA GAEHWILER, BRIAN GROSS, THOMAS KÖLLNER,
WENDY PROCTOR, AND DAVID ZILBERMAN

INTRODUCTION

One of the major tools for improving environmental quality is the implementation of a PES scheme. Such schemes include government programmes, such as the conservation reserve programme (CRPs) in the United States, forest conservation for debt arrangements reached between rich developed countries and tropical developing countries, and purchases of land to become nature reserves by The Nature Conservancy.

While there are plenty of offers and schemes to supply ecosystem services (ES), the existence of the necessary demand for such schemes is less obvious. Some of the prominent PES arrangements are also transfer payments by governments to constituencies such as farmers. While there may be much stated WTP for ES, some of which is probably unsubstantiated, actual WTP is significantly smaller. For PES to join 'command and control', 'polluter-pays', and 'cap and trade' schemes as a major tool to achieve environmental objectives, procedures to create and utilize the demand for ES must be established. The objective of this chapter is to study the challenges of marketing ES, and to develop a framework that could assist in marketing such services effectively. Marketing efforts should consist of identifying the various market segments and their needs, developing products that are user-friendly and straightforward, developing marketing tools to inform potential buyers about the product, and reducing concern about performance, reliability, and fit. Thus, following the introductory section where we present the various categories of ES, we will identify the different categories of potential buyers of ES, the factors enhancing their valuation of ES and WTP, and the product design and marketing strategies necessary to establish significant effective demand.

To investigate the level of marketing performance of selected suppliers in Latin America, a questionnaire was sent to 49 different organizations providing ES. Nine of them replied—timber companies, NGOs, governmental organizations (GOs), and transaction managers. The results of the questionnaire study showed that marketing plays, as yet, only a minor role in business planning and so does the evaluation of the demanders' needs. The customers' needs and expectations are hardly known and hence the products are not accordingly designed. Although there are some exceptions, it can be stated that marketing strategies are virtually non-existent. Often, marketing is reduced to mere promotion. We conclude that improved marketing, understood in a comprehensive way, can help the supply side to address the potential demanders' needs and increase market transactions.

THE DIVERSITY OF ES

ES serve several purposes. PES are used as incentives for pollution control, resource conservation, and provision of environmental amenities. Notable examples of pollution control are arrangements through which water utilities pay dairy farmers not to use grazing practices in the watershed of their reservoir. Conservation is achieved, for example, by payments to forest communities to control deforestation (Alix-Garcia et al. 2004). Wetlands and wildlife refuges are being created to provide recreational opportunities for hunters and bird watchers, and to provide water purification as well as buffering for flood control for nearby cities.

Control of pollution can be achieved by several means. Command and control schemes[1] have been heavily used by regulators, but there is vast evidence showing that they may lead to inefficient outcomes. The polluter-pays principle,[2] as embodied by a Pigovian tax, has been long advocated by economists and environmental activists alike

---

[1]Command and control instruments (for example, mechanisms, laws, and measures) rely on prescribing rules and standards and using sanctions to enforce compliance with them.

[2]Where the polluting party pays for the damage that they have inflicted on the environment.

as an efficient and effective policy. However, as Buchanan and Tullock (1975) argue, polluters may use their political muscle to prevent imposition of these policies. Cap and trade[3] schemes are another mechanism that can lead to efficient outcomes and have been widely used in recent years, for example, in the Kyoto Protocol. They may be more accepted by polluters, because unlike pollution taxation, they do not withdraw resources (tax payments) outside of the polluting sector. Even given taxation, polluters may yet prefer not to spend resources on pollution prevention, or they may not be able to afford to do so. PES can be a subsidy to reduce pollution, and it may occur when polluters are politically strong and have rights to pollute, or when they are very poor and do not have the resources to pay for pollution prevention. Similarly, payments for conservation may occur in situations where the self-interests of the resource owners conflict with those of a third party wanting the resource preserved. For example, forest communities in developing countries may have high discount rates and are likely to perceive large gains from conversion of forest resources to rangeland and lumber, while environmentalists and natural resource agencies may prefer to see the forests preserved. Growing demand for eco-tourism, for example, safaris, bird-watching tours, or hunting, may lead to payments to landowners and developers to preserve or create the environment that provides these specific amenities coveted by consumers and recreationists.

Another way to distinguish between various ES is to separate those providing consumptive use from those providing non-consumptive use. The case of a soft drink company paying farmers to divert waste disposal to improve water quality as well as that of a recreational club paying to preserve hunting grounds are examples of payments for consumptive use. Yet, many individuals may pay for non-consumptive use as well. Some ES represent the existence value of knowing that a rare species will survive or a pristine environment will remain unchanged.

Many ES, for example, sequestered carbon in the soil, trees, or other media, are becoming standard commodities treated by large markets. Others are more unique, and more resistant to commodification. In

---

[3]Where a central authority sets an upper limit or cap on the environmental impact to be achieved (for example, the amount of a pollutant that can be emitted) and individuals or groups involved in the scheme are given credits to define the amount that they are allowed to each contribute to the total impact. These groups can then trade these credits as they wish, thus ensuring that the overall limit is not exceeded.

some cases, PES are a one-of-a-kind experience or for the continued existence of a unique natural phenomenon.

The same land or water resource may provide more than one ES. Diversion of land from intensive farming to forest may improve air quality and protect against soil degradation. A wetland may provide both water purification and wildlife habitats. In these cases, both complementary activities can provide income that will allow for greener activities. In some cases, ES obtained from a resource may be substitutes. A piece of land may be conserved to provide forest services or may be diverted to become a wetland.

Like many assets, both timing and location impact the use and valuation of ES. A well-maintained vineyard may have a much higher value as a source of recreation and aesthetic beauty in the urban fringe than it would in the agricultural heartland. A water reservoir carries a much higher value during a drought than during the rainy season. The opposite is true for a wetland acting as a flood control buffer, which may be especially valuable during periods of heavy rain.

The spatial dimensions of resources providing ES vary necessarily, often depending on biological considerations. Some environmental amenities exhibit increasing returns to scale. For example, a critical mass of land resources may be needed to sustain certain wildlife species. In other cases, there may be gains from maintaining land or water resources spread over separate locations for protection and diversification. Some protected species may be spread over a larger area but must have open pathways to move from place to place.

The multiple typologies of ES and their benefits are useful and necessary in designing and targeting marketing strategies and identifying potential buyers, sales channels, mechanisms for exchange, and information awareness and promotion activities.

THE NATURE OF DEMAND FOR ES

The development of a marketing strategy for ES is derived from understanding the demand for it. As we have seen, ES are diverse and are likely to be purchased by diverse buyers. The considerations affecting some of the buyers are analysed below.

Governments

Most of the spending on ES thus far has been done by government agencies at several levels—local, regional, national, and international.

As Rausser (1982) notes, some government policies aim to address market failure and others are distributional. That holds true for ES programmes. The CRP in the US has served for a large part as a farm subsidy programme, and the policy debate surrounding its design is about aligning environmental criteria with the political power of affected parties (Babcock et al. 1996). The environmental quality improvement programme (EQIP) is paying farmers to reduce pesticide use and livestock producers to reduce animal waste. It combines pollution control with transfer payment. The payment for forest conservation in Costa Rica is another example where the ES are combining transfer of resources together with attaining environmental objectives. The Global Environmental Facility is an example of a mechanism for provision of funding for ES by a 'global' government.

The allocation of funding for ES by governments can be modelled by a cooperative game framework similar to the one proposed by Zusman (1976). Let i be an indicator of ES projects, assume that the funding of 1 projects is considered, where $i = 1, \cdots 1$, and let x_i be the budget allocated to the ith ES project. Assume that there are K groups affected by these projects (positively and negatively), and the benefit the kth group obtains from the ith project is $B(x_i, k)$. Each of the groups can affect the political process (lobbying, for example), and the political weight of the kth group is denoted by $w(k)$. We also assume that there is a political cost associated with spending given by $c(x)$ where $X = \sum_{i=1}^{1} x_1$. With these definitions, the budget will be allocated to the various ES projects to maximize a weighted sum of benefits minus the total political cost of the budgetary expense. The optimization problem, solving for the various x_i, is thus:

$$\underset{x_1, x_2, \cdots}{MAX} \sum_{k=1}^{K} \sum_{i-1}^{I} w_k B(x_i, k) - c\left(\sum_{i-1}^{I} x_i \right) \qquad ...(8.1)$$

The expenditure on the ith ES is determined by equating the sum of the politically weighted marginal benefits of that ES with the political marginal cost of expenditure. Groups with more political muscle, or groups that are ready to sacrifice more political capital on garnering ES projects that benefit them will bias the allocation of ES money to favour their projects. Thus, agricultural groups in the United States and forest owners in Costa Rica may obtain significant support to ES that benefits them disproportionately, because of their political investment. The condition also suggests that when the government finances are in better shape and the cost of expenses is lower, more will be spent on ES.

Governments can also induce demand for ES indirectly. Stricter water quality standards, restrictions on chemical use, and tougher land-use regulation are activities that may drive the creation of PES schemes by affected firms. Higher water quality standards may induce water utilities to pay landowners in their water catchment areas to modify their activities. The imposition of constraints on wetlands conversion in the US led to the evolution of wetland banking. The inclusion of carbon sequestration activities as a means to obtain carbon credits, consistent with the Kyoto Protocol, is creating new ES-generating activities. The regulatory process has been a major avenue for creating demand for ES, and indeed environmental groups have been working through the political process to introduce policies that directly or indirectly lead to ES programmes. Such political activities may require establishing alliances with various groups that sometimes may be adversarial to environmental causes. Indeed, farmers or forest landowners are likely to work together to garner payments for conservation activities, despite differences on major issues. Marketing of new PES requires political action, persuasion, and coalition building to obtain the political support for policies that directly or indirectly induce the emergence of either public or private demand for ES.

Industry

Most private firms will view payments for ES primarily from a profitability perspective. The weighting of risks, short-term vs long-term earnings, and the goodwill generated because of environmental stewardship may vary among firms and affect their demand for ES, but fundamentally, the demand for ES is derived from the economics of their main business activities. Some firms may pay for ES to improve the quality of production inputs, other firms will pay for ES to reduce risks, still others may pay for ES to help them improve their image, thus enhancing the demand for their final product.

Consider first the demand of a bottled water (or soft drink) industry for ES that will enhance the quality of water it provides. For simplicity, we assume that the industry is competitive, but the analysis can be easily modified for a monopolistic or oligopolistic one without substantially changing the implications.

Suppose that one can have either high quality or low quality water. The quality of the water is affected by agricultural production activities. One way to improve water quality is to pay farmers to modify their production technology. Let the derived demand of the product with high quality water, when the consumers are informed, be denoted by $D(p, h)$ in Figure 8.1, and the demand for the product with low quality water be denoted by $D(p, l)$ in Figure 8.1. The marginal production cost of the product with water of either quality is MC. If the higher quality water is utilized, the price is P_n, the quantity consumed is q_n, and the industry's profit is the area ABC. If the lower quality water is used, the price of the product is P_l, the quantity consumed is q_l, and the profit is the area AEF. The increase in the industry gross profits from the provision of ES that will increase water quality if the consumers are aware of the quality issue is denoted by MG and is represented by the area FEBC in Figure 8.1. This area is the maximum amount that the industry (represented by a producer association) may be willing to pay farmers for the ES. Of course, the industry would like to keep some of the extra profits, and the final distribution of profits will be the result of a negotiation. Let the payment for ES be denoted by PES. If the cost to the farmers for modification of their production system is FC, then $Fc \leq PES \leq MG$. If the negotiation results in a fair solution (Rabin 1998) where

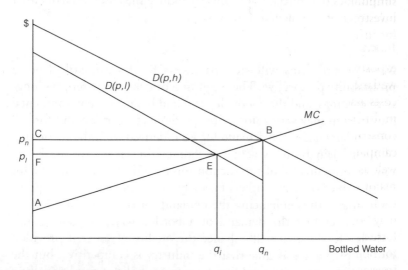

Fig. 8.1:   Demand for Bottled Water with Low and High Water Input

the gain is shared, then $PES = (MG - FC)/2$, but that is one among many possible outcomes.

Consider the case where the clients of an insurance company are exposed to flood risk. Both the probability of a flood event and damage it causes depend on the size of a wetland (for example, buffer zone) in the flood plain. Let $q(W)$ denote the probability of a flood, and let $D(W)$ be the total cost of the damage caused by a flood for a buffer zone of size W. Let $c(W)$ denote the cost to implement and maintain the wetland. A risk-neutral firm will determine W to minimize expected total cost:

$$\min \{q(W)D(W) + c(W)\}$$

The optimal size of the wetland will equate the expected value of the marginal savings with the marginal cost of additional wetland capacity. Both the liability and expectation of a flood can change over time, revising the calculation and WTP for ES (wetlands). For example, as a city grows, the value of a wetland insulated from flood risks increases as a result of the increasing total property values at risk. Therefore, we may expect insurance companies to increase their investment in risk-reducing ES over time. The analysis should be expanded to cases where the insurers are risk averse, and where the levels of liability, risk premium, and wetland capacity are determined simultaneously. It would be useful to investigate situations where investment in ES may reduce risk premiums and increase profitability for insurers simultaneously.

Frequently, insurance is competitive, and several insurers bear exposure within a given flood zone. Thus, raising funds to support wetlands or other ES to reduce risk may entail high transaction costs associated with collective action among the companies, or with multi-party negotiations with the agency associated with wetlands construction and preservation. In these cases, designing a marketing campaign to raise awareness about the potential gains from ES, as well as building the goodwill to allow financing of the project, is a major challenge. In some cases, re-insurers could be better prospects for financing risk-reducing ES, because they are small in number and may be the ultimate bearers of the risk (in the case of large catastrophic losses). Since the Government acts as the insurer of the last resort for flood risk in the US and other countries, it would be worthwhile to examine the role of governments as purchasers of ES to mitigate this risk, or to consider a public–private partnership to this end.

Firms may use payments for ES as a mechanism to enhance public image. This is likely to happen when consumer preferences, and therefore demand, depend on the environmental record of the firm. Let the demand for a firm's product x be affected by ES expenditures; so if x is the quantity of output, and p is the price, then demand will be $x =D(p, ES)$.

Let us suppose that the cost of production is $c(x)$, and the cost of ES is $v(ES)$. Then the firm has to decide on optimal output and optimal spending on ES. The optimal spending on ES is determined where the marginal increase in revenue because of ES is equal to the marginal cost of ES:

$$\frac{\partial R(x^*, ES^*)}{\partial ES} = \frac{\partial v(ES^*)}{\partial ES}$$

Revenues will increase from spending on ES when synergies exist between the product and ES. For example, buyers of outdoor recreation gear will value a company more that actively supports the preservation of pristine wilderness. In some cases, companies with a bad environmental reputation may use PES to soften their image and recapture some environmentally conscious consumers. In these cases, the benefit from PES can go beyond increased sales, and may reduce the fervour that leads to expensive lawsuits or restrictive legislation.

Consumers

Consumer demand for ES programmes will be expressed by paying for the output of these programmes, for example, entrance fees for a wetland or a park, purchasing hunting or fishing licences, or buying products that are produced by endogenous tropical communities living in an ecologically sustainable manner within the forest. Another form of consumer demand that may be more intriguing is contributions to NGOs or other organizations for funds generating ES and preserving forests, wildlife habitats, and wetlands. Indeed 'angel investors' that provide significant sums to environmental causes have played significant roles in financing green activities that include provision of ES.[4] Our analysis will concentrate mostly on contributions that fund ES programmes.

[4]See, for example, Investor's Circle, a group that coordinates the efforts of environmentally conscious angel investors for sustainable business: *www.investorscircle.net*

Individuals will pay for ES because they either directly benefit or indirectly gain from existence value. Hunters and bird watchers will contribute to support the establishment of a local wetland that will enhance the value of the local ES they consume. However, some individuals may support an ES programme that is farther away because it generates valuable outputs that are global public goods, it contributes to biodiversity preservation, or they value the existence of an amenity. Another element exists as well—consumers may contribute for their reputation effects or to gain social status as donors. Many public goods such as symphonies, theatres, and museums thrive on donations from individuals who derive social standing or out of a sense of social obligation or citizenship. The combination of these incentives should drive marketing strategies that aim to raise funds for ES projects.

Let an individual's willingness to contribute to an ES fund be denoted by $V(W, A, S, t)$ where W is wealth, A is awareness and experience, S is social pressure, and t is incentives. Contributions to causes are likely to be luxury goods, and the marginal contribution is likely to be positive and increasing with wealth ($\partial V/\partial W > 0$ and $\partial^2 V/\partial W^2 > 0$). However, wealth by itself is not a sufficient condition for contributing to an ES fund. People are more likely to contribute when they are familiar with the causes and care about them ($\partial V/\partial A > 0$). Moreover, people who care about the causes and can afford to pay have the best potential as valuable contributors ($\partial^2 V/\partial A\partial W > 0$). Contributing to a cause may generate social externalities and reputation effects, implying that people are more likely to contribute to causes supported by their peers or social group ($\partial V/\partial S > 0$). Finally, contribution to ES programmes and other causes can be induced by incentives such as tax exemptions ($\partial V/\partial t > 0$), and with a progressive income tax, the tax savings is likely to be more significant for wealthier individuals ($\partial^2 V/\partial W\partial t > 0$). Furthermore, naming opportunities are valuable incentives, especially when they are appreciated by the donor's peer group ($\partial^2 V/\partial S\partial t > 0$).

Thus, a marketing effort to increase consumer contributions to ES funds has to first target a group of potential donors who combine income and interest. Interest and awareness can be generated by various informational policies, including advertisements, informationals, and events to expose potential donors to environmental causes and their benefits. The Sierra Club organizes outings both to generate awareness of environmental causes and to build social networks that are likely to enhance ES fund contributions. The Sierra Club, and other environmental groups provide products that bundle 'fun' (such as adventure trips) with efforts to support a cause. This synergetic product has wide appeal. Awareness of environmental causes can also be enhanced by improving environmental education in schools and in colleges, and by 'teaching the teachers'. Education and awareness of ES are not only beneficial for fundraising, but they are also beneficial for building political support for public provision of ES. Finally, garnering public support for ES programmes requires establishing an incentive structure that includes tax deductions, as well as various forms of recognition. Universities may name buildings after important donors, and cities may name streets and public plazas after politicians. Similarly, a contribution to establish ES could lead to other naming opportunities.

The heterogeneity of the public suggests that efforts to create public demand and contribution should adjust according to the situation. Some people who may not be familiar with a problem or concept behind the ES should be targeted by basic educational programmes and other awareness-raising efforts. Individuals who are more familiar with the issue should be targeted more intensively as potential donors. Of course, some especially promising prospects may deserve special attention. Past donations reveal preferences for the ES, thus a record of donations in the past can be used as a basis for pursuit of future contributions.

NGOs

Frequently, NGOs are formed to articulate the preferences of economic agent groups engaged in activities on their behalf. When society consists of heterogeneous consumers, there may be a group of individuals in society that will have stronger preferences than the average citizens for a public good, say, the preservation of certain ecosystems. The members of the group may perceive underinvestment by the government in the preservation of the ecosystem, and raise the funds to both invest in the preservation directly and to lobby for enhanced public support. The WWF initiates and supports activities that preserve and protect wildlife using donations from individuals who have strong preferences for wildlife protection and improved well-being. NGOs may pay for ES as part of their activities or use their political muscle to enhance government payments for

ES. The Nature Conservancy has adopted payment for ES as a major element of its mode of operations. Other environmental NGOs may emphasize the role of penalties and direct control to achieve environmental goals. That may be because of either lack of capacity to raise funds for PES, lack of expertise in their design and use, and objection to using these forms of financial incentives. PES is a relatively new institutional innovation, and it has been only partially adopted by organizations that can benefit from it, in particular, NGOs. As the adoption models suggest (Feder et al. 1985), the adoption of PES may increase over time as a result of imitation among NGOs, education and training of NGO personnel, and accumulation of experience that will lead to reducing the cost and increasing the effectiveness of PES programmes.

As we analyse demand for ES, it is useful to differentiate between mechanisms and institutions that are responsible for funding ES, and the institutions that are responsible for the actual purchase transaction with the seller. NGOs tend to be organizations that buy ES, but the funding of the purchases is generally supplied by consumers. Industries may, for the most part, be direct buyers of ES (even if they make direct donations to NGOs), and governments in general are direct buyers of ES.

DESIGNING PRODUCTS BASED ON ES

Because ES are basically processes and not physical objects that can be directly experienced, it is essential in marketing to design products that have a material existence to potential clients. In ES marketing, tangible elements are combined with intangible elements in order to achieve a reality experienced by the client (Shostack 1977). The core goal of product design for services then is to 'package' a service with its tangible characteristics, focusing on the evidence or clues that clients use as indication of the success or performance of the service.

Monitoring and evaluation are, therefore, crucial components to ES marketing, because it is part of the product development. In cases with clear indicators for outcome that can be effectively measured, the product can be designed around such performance measures. For example, a river restoration project to improve water quality can be defined based on the expected changes in nutrient concentration and fish populations.

As is often the case for ecosystem management, however, there are no effective measures of outcome. A second best, therefore, is to define the service in terms of the actions aimed towards the outcome. This is true of pollution control and conservation, which are most often defined in terms of pollution *abated* and land uses *not* employed rather than their actual respective benefits. In both cases, the service is essentially an agreement, and marketing efforts can be directed towards enhancing the tangibility and security of such agreements with certificates, contracts, and assurances.

Table 8.1 shows the potential and existing products for ES in general and three specific ES (carbon sequestration, watershed protection, and biodiversity protection), their customer groups, the type of benefit generated, and the associated payment type. In this table we focus only on pure ES, ignoring those that are bundled with physical products, for example, certified timber, organic agricultural products.

The payment mechanisms are associated with the type of benefit expected. For an investment the 'buyer' clearly expects a direct financial benefit, and for a donation a non-financial benefit. For example, a private donation to a conservation oriented-NGO may be defined in terms of increased environmental utility, whereas a company investing in forest protection for improved water quality and/or company image expects a (probabilistic) financial return. Another form of payment can be defined as compensation: a financial or non-financial benefit to offset environmental impacts. Finally, a subsidy is a payment by GOs in order to achieve non-financial benefits.

Table 8.1: Possible Products and Types of Payments for Ecosystem Services

Service	Product	Customer group	Type of benefit	Payment type
Ecosystem Services in General	Venture capital fund specialized on organizations producing 'sustainable' goods and services	Investors	Direct financial benefit	Investment
	Micro credit fund for forest owners providing ecosystem services	Investors	Direct financial benefit	Credit
	Subsidies for ecosystem services, generated mainly from taxes	GOs	Non-financial benefit	Subsidy

(contd...)

Table 8.1 (contd...)

Service	Product	Customer group	Type of benefit	Payment type
Carbon Sequestration	Compulsory Compensation through Certified Emission Reductions (CER) in the framework of the Kyoto protocol	Companies participating in the Kyoto protocol	Indirect financial benefit	Compensation
	CER+ for carbon sequestration with additional benefits in biodiversity and rural development	Companies participating in the Kyoto protocol	Indirect financial benefit	Compensation
	Voluntary compensation through verified emissions reduction carbon-offset credits	Universities, schools, communities, consumers	Non-financial benefit	Compensation
	Tradeable carbon credits	Companies	Non-financial benefit	Compensation
	Funds for carbon sequestration or other land use projects to off-set carbon emissions	Companies	Indirect financial benefit	Compensation
Watershed Protection	Ecosystem service certificates for watershed protection provided by forests	Local companies and individuals consuming water (hydropower, drinking water, agriculture)	Direct financial benefit	Investment
Biodiversity protection	Biodiversity certificates: ensure protection of biodiversity rich land and are tax-deductible	(Inter) national companies, NGOs	Indirect financial benefit	Donation
	Permission for bioprospecting or research	Pharmaceutical industry, universities	Direct financial benefit	Investment
	Remediation fund to compensate land use impacts, other negative impacts on biodiversity	Mining companies, contaminators	Indirect financial benefit	Compensation

Source: Authors.

INSTITUTIONAL ARRANGEMENTS FOR FUNDING AND PURCHASING ES

Introduction of PES requires the establishment of an institutional set-up that will allow sellers to discover buyers, provide an environment for negotiations about price and the details of the ES, and offer mechanisms for protection against various risks. When the ultimate buyer is different from the provider of funding, then there is a need for institutions to raise funds. The exact set-up depends on the number and identity of buyers and sellers, and the features of the ES.

A market embodied in an exchange is likely to emerge as a mechanism where ES are bought and sold in cases where the number of buyers and sellers are large, and the ES can be standardized and commodified. The markets for carbon emission rights are obvious examples as sequestration or emission of 1 tonne of carbon is a well-defined product, and in this case location of emission or sequestration does not matter. But obviously some means of sequestration are more certain and stable than others, and implementation of these markets requires a system of certification. When there are many buyers and sellers who are concerned with issues of liquidity and certainty, in addition to the spot markets in the ES, markets for options as well as future markets may emerge.

Sometimes, when there are many buyers and sellers, but the products are very diverse and non-standard, many of the transactions will take place through a bulletin board. Buyers describe their product, possibly showing a picture and providing contact information, or alternatively, the sellers define what they want and provide contact information. In this case, the negotiations between buyers and sellers establish the price and the specifics of the transaction. A bulletin board is very popular in many important markets, for example, used cars, dating, and jobs. A special example with many similarities to ES markets is the real estate market, where every house has its own unique attributes. There is an electronic water market for the California Central Valley, where buyers and sellers use the Internet to meet and arrange for transactions (Olmstead et al. 1997). Trading through a bulletin board requires higher transaction costs than trading through an exchange, and market information within a bulletin board system is not as transparent, but these extra costs are appropriate if buyers and sellers need to adjust for the uniqueness of products. When there is one buyer (or a small number of buyers) and many sellers, then the purchase of ES will be managed through bidding. For example, in the case of the CRP programme in the US, the government asks farmers to provide a bid regarding the amount of land that they will offer to the programme, the annual amount of money they will demand, and the environmental benefits of the land. The government determines the sellers of ES based on a formula that weighs the costs and benefits of different proposals. A similar system of bidding exists in government programmes in Mexico and Costa Rica that pay for ES produced by forests and in Australia to encourage the growth of native vegetation (see Box 8.1).

A seller of ES may choose to open a storefront, namely, they may provide a specific address or location where people can purchase ES. Storefronts are opened when there are relatively many buyers and a small number of sellers, and the buyers need the opportunity to examine and compare a variety of products in order to find the one that fits their needs. In the case of wetland banking, specialized organizations have an inventory of wetlands that can provide the appropriate services to developers who need wetland services credit in order to develop their own land. A storefront can be a physical location, or it can be a virtual storefront online or via catalogue. Potential eco-tourists choosing a locale providing recreational or

---

Box 8.1: An auction scheme—Bush Tender

BushTender is a relatively new approach in Australia offering landholders the opportunity to receive payment for entering into agreements to provide management services that improve the quality or extent of native vegetation on their land. These services are based on management commitments over and above those required by current obligations and legislation.

Landholders choose their own auction price for the management services they are prepared to offer to improve their native vegetation. This price forms the basis for their bid, which is compared with the bids from all other landholders participating in the process. The proposed actions are compared using a 'biodiversity benefits' score that was calculated for the farmers by BushTender officials in pre-bid site visits. The successful bids are those that offer the best value for money in terms of the highest biodiversity benefits score for the lowest offered price.

scenic ES may visit ranger stations or shop virtually through brochures, or national park websites. When storefronts exist and buyers of ES have a choice between different types of products, then the role of marketing tools such as advertisements and pre-purchase demonstrations become particularly important. Mechanisms to reduce buyer uncertainty about the product may increase demand and may increase sales.

When it comes to raising funds for ES, a major mechanism is soliciting through various forms of fundraising. The potential donors are heterogeneous; they vary in their preferences and in their ability to pay. Furthermore, the fundraising organization may have information about some subset and may be uncertain about another. Some forms of solicitation through advertisements and general media are used to both identify individuals who may have interest in becoming donors for the purchase of ES, and also to compel them to contribute. A letter campaign may target individuals who are known to have preferences for certain types of ES (members of certain groups or associations). For richer individuals, the organization can afford solicitations in person.

ENHANCING THE DEMAND FOR ES

The diversity of situations that gives rise to ES, leads to the use of different channels and mechanisms to fund and sell ES. The demand for all the diverse categories for ES is likely to increase if several basic principles are recognized. Consumers are concerned about risk and, in particular, risk about product reliability and fit (Heiman et al. 2001). Buyers are not likely to purchase a product that has a high likelihood of failure, and mechanisms such as warranties and dealer backups have been used to improve the performance of a product and reduce the cost of failure for the consumer, thus increasing demand. Similarly, consumers are less likely to buy a product that may not fit their particular needs. Mechanisms such as product demonstration, money-back guarantees, and secondary markets have been used to reduce fit uncertainty.

In the case of ES, reliability is of utmost importance. When a utility in the North buys carbon credit from farmers or forest communities that are to engage in carbon sequestration, the utility is concerned about the possibility of the other party failing to fulfill their commitment. Similarly, when a water utility pays livestock operators to control animal waste run-off, they may be concerned about violations. The concern with risk of violations or failure to fulfill contracts may reduce the willingness to pay for ES in both cases. One solution is to develop effective mechanisms for monitoring the behaviour of providers of ES, and for enforcement of contracts. This is quite a challenge, especially when many small farmers are providing carbon sequestration, or are committed to reducing certain forms of polluting behaviour.

Thus, creative mechanisms that reduce the cost of monitoring and reduce the likelihood of violations are of significant value. One example is the use of collective punishments as a threat, and another is reduced penalties as compensation for self-reporting when accidents occur, in order to reduce the damage.

It is much easier to assure delivery of ES when paying on an ongoing basis than to assure delivery of future ES from a purchased asset. Some ES programmes pay the provider annual rent for access to services. For example, the CRS of the US, as well as the forest protection programmes in several developing countries, pay landowners on an annual basis for the choices that provide the ES. In other situations, the buyer purchases an asset, which is assumed to provide the ES each period during its lifetime. For example, a programme to purchase rainforest acreage pays for the ES associated with that acreage indefinitely into the future. Similarly, when a farmer is paid for sequestration of a tonne of carbon by using no-tillage or other technologies to reduce carbon emissions, it is assumed these choices will not be reversed in the future. Establishing reliable mechanisms of monitoring and enforcement, as well as insurance, is a bigger challenge when one plans the long-term sale of ES-generating assets. Providers of ES may offer ES payments of varying durations, and adjust prices for uncertainty costs of insurance and the discount rates.

Buyers of ES may be subjected to fit risk as well. For example, a utility may pay farmers or invest in a wetland protection to improve water quality, and then realize that these activities did not solve the water quality problems. Contracts that provide the buyers some compensation or refund in the event of the ES failing to serve its intended purpose are likely to increase the expected net earning from the provision of the ES, as buyers will be willing to pay more for the ES when their fit risks are smaller.

In addition to the buyers' preferences for lower risks, they also value that the buying process be less costly in terms of time and money. Buyers need to spend less time purchasing a product if there are fewer uncertainties about its performance, use, and price, and when the purchasing process is streamlined and simple. One avenue to reduce the costs and risks of purchasing is selling standard commodities. One of the challenges in marketing ES is commodification, namely, established uniform product standards that allow large-scale exchange. When ES that remove pollution and waste can be sold as commodities, they can be incorporated into cap and trade arrangements. This is the case for carbon sequestration activities that have been incorporated in the trading of market emission rights. Wetlands that reduce the nitrate loading of a body of water through bio-remediation may be paid for through PES, once pricing for removal of the nitrates in the water is established through cap and trade arrangements or pollution penalties.

When an activity is providing a mix of ES, if each ES has a market- or policy-determined price, the activity will be paid by the sum of values it creates. In other cases, weights will be given to the various ES resulting from the activity, and markets or other mechanisms will determine the price of 'standard' units of ES that will be used as a benchmark for assessment. This approach has been used in assessing bids or determining payments for participation in several governmental ES programmes in the US, Europe, and Mexico.

Buyers of ES, especially private buyers, may also be concerned about liquidity. They would like to be able to re-sell the product if their preferences or financial situations change. A related concern of buyers and sellers of products is price variability and uncertainty. Therefore, futures and options markets have evolved, allowing hedging of prices into the future. Active exchanges for well-defined commodities serve to enhance liquidity, and are especially effective if futures and options markets are included because they reduce the risk of market participation. Indeed, there are some active environmental and resource markets that have futures and options; they include the water bank in California, which serves as a mechanism to protect water districts and farmers against drought. The exchanges for tradeable permits for air pollutants (including the emerging market for carbon emission rights) also combine spot, futures, and options markets.

The pursuit of lower transaction costs and uniformity has to be balanced against the gains of specialization. Buyers of ES are diverse in their preferences and ability to pay. Providers of ES should aim to establish quantitative standards that will allow for easy value assessment and trading of ES, yet at the same time allow for differentiation among them. Differences in terms of size, quality, and other dimensions that can be easily monitored allow buyers the opportunity to be selective and enable the seller to take advantage of various categories of buyers interested in the ES. A forest community may gain from payments for improved water quality by a nearby utility, earnings from carbon credit sold to a global carbon exchange, and premiums for wildlife protection paid by a conservation group.

The gains from differentiation are not limited to the sale of actual ES; they can also be present in fundraising by environmental groups planning to purchase ES. Many donors would like recognition by having a location named after them, receiving special acknowledegment (for example, in a newsletter or other publication), or being taken for a personal visit to exotic sites. Differences in giving capacity and interests among donors can be accommodated by offering a variety of naming opportunities, establishing different categories of donations that are related to contribution size, or by organizing special and exclusive environmental tourism or 'adventure in nature' programmes. This type of entrepreneurship will enable an environmental group to obtain more funding for its activities.

An alternative approach to raising funds and support for ES is through appropriate development projects. Tourists who come to a resort may appreciate nature adventure around it and thus tourists visiting recreational facilities may provide the funds that support provision or maintenance of ES in the vicinity of these facilities. Again, price discrimination that increases the surplus taken by the facility, and therefore by the environmental group for purchase of ES, can be a viable strategy. For example, allowing the resort to build both luxury and regular rooms in a hotel near an animal reservation will generate higher earnings for the facility than having standard rooms only; and, if a given fraction of the income ends up as PES, the higher earning of the hotel with differentiated rooms will result in higher ES.

MARKETING SURVEY OF ORGANIZATIONS PROVIDING ES FROM TROPICAL FORESTS

In order to investigate the existing state of marketing for ES, an inquiry was made in 2004 about the activities on the supply side. This was done by means of a questionnaire, whose aim was to assess the quality of the marketing measures taken. This questionnaire was structured into three parts: general information, financial information, and a third part that actually assessed the marketing activities with respect to: (i) product development, (ii) financing mechanisms, as well as (iii) promotion and advertisement. In all, 49 organizations located in Latin America were contacted and 9 of them provided a complete questionnaire (18 per cent response rate). The organizations, which responded, were active in Costa Rica, Nicaragua, Chile, and Colombia. Their number of employees ranged from 3 to 6448 (median 39). All of them possess predominantly monoculture forest plantations and are privately organized (four collaborate closely with governments). In the following subsection we present the results of the questionnaire study.

Product Development

As different types of services provide different benefits, it is important to know the needs of customers and take those into account in product design. For that reason the questionnaire asked if the most important needs of the customers and the main benefits of the services provided for the customer are investigated and known. Only a few organizations systematically investigate the needs of their potential customers (n = 2). Most of the organizations that responded base their product primarily on their own experience or on the experience of other supplying organizations. The ES that is most commonly provided is carbon sequestration (n = 9) and its benefits seem to be most obvious to the project owners. It is offered by all the projects. The most commonly sold product is the certificate that attests the voluntary compensation of carbon dioxide emission. In some cases they also plan to sell CER in the framework of the Kyoto protocol. The respondents mentioned that the benefit for the customer in case of carbon sequestration is compliance with Kyoto rules, improved image, and even good feeling, because the personal carbon emission is offset. For biodiversity protection (n = 7), the respondents mention the following benefits for the buyer: compensation of negative impacts like land use, improvement of image, and investment opportunities. The only benefit of watershed protection (n = 5) that is mentioned is that high quality water is supplied. Interestingly, image benefits are not associated with watershed protection.

Financing Mechanisms

The financing mechanisms of the different respondents supplying ES vary significantly. More than half of the participants use more than one strategy of funding the provisioning of ES. The source is often a mixture between public and private funds. The most important financial resources are donations and investments. The main donors are foundations and individuals. Table 8.2 shows the distribution of financing mechanisms.

Table 8.2:   Distribution of Financing Mechanisms

Participant ID	Source of financing in per cent						
	Investment	Credits	Sales (e.g. timber)	Donations	Subsidies	Other	Total
1	0	0	0	100	0	0	100
2	0	0	33	33	0	33	100
3	0	0	0	0	0	100	100
4	0	0	100	0	0	0	100
5	20	40	20	20	0	0	100
6	5	0	0	5	90	0	100
7	20	20	40	0	20	0	100
8	100	0	0	0	0	0	100
9	50	0	0	30	0	20	100
Mean	22	7	21	21	12	7	100

Source: Authors.

Pricing ecosystem services per unit is a key element of marketing. The most common ways to establish a price are using a value fixed by an administration (n = 3) or the one suggested by supply and demand as the outcome of a negotiation (n = 3). In other cases, prices were supply driven, that is, marginal costs of the service were calculated and the price set according to this value (n = 2), or demand driven, that is, based on WTP (n = 1). One organization mentioned equitable benefit sharing as a price-setting principle. The number of answers does not add up to nine because multiple answers were permitted and two organizations use more than one price setting strategy whereas another left the question blank. Four organizations have conducted studies to evaluate the WTP of their customers and seven out of nine think it would be very important (seven points out of seven) to know this value. Of the remaining two participants, one did not estimate the same to be very important and rated it with three points, and the last one did not answer the question. Although the WTP was rated as very important in the questionnaire study, it appears only once in the list of price setting strategies.

Promotion and Advertisement

The questionnaire study also revealed the measures related to promotion and advertisement taken by the organizations supplying ES based on tropical forests. The investigation showed that the majority of services were sold directly to the customer and no intermediary was involved (n = 5). The type of transaction depends on the location of the customer and this fact is linked with the type of service. For example, water services are usually sold to consumers within the watershed, such as breweries or hydropower plants. Also a combination of direct and indirect distribution was used (n = 2), but only indirect distribution was also used through intermediaries (n = 2).

We also focussed on promotional measures taken by the participants. They assessed the effectiveness e of the measures they used to gain new customers. The scale was from 1 (not effective) to 7 (very effective). To establish the ranking, the number of answers n of a certain measure M were multiplied with their corresponding effectiveness e.[5] Some participants stated that they had applied certain measures but failed to rate them. These 'unrated' measures are given the lowest rate 1. The total rating was calculated as $M \sum_{i=1}^{n} e_i * n_i$. The results presented in Table 8.3 show that direct promotion was perceived as more important compared to more indirect advertisement. Highest rating and ranking were obtained by activities to convince

[5]Effectiveness is referred to here in terms of the ability of the implemented measure to help in selling more ecosystem services.

Table 8.3: Rating M and Ranking of Promotion Measures Taken by Participants

		Rating of effectiveness e from 1 to 7 (times named by participants in cells n)									Total rating M
		1	2	3	4	5	6	7	not rated	0 not used	
Rank	*Direct Promotion*										
1	Convince people directly					1	1	5	1	1	47
2	Activities within organization					1	1	3	1	3	33
2	Activities for local people					1	1	3	1	3	33
3	Public presentations abroad					2	1	1	1	4	24
4	Personal letters	1	1	1		1			1	4	16
5	E-mail		1			1	1			7	14
6	Excursions						1		8	7	
6	Invitations						1		8	7	
7	'Round table' activities					1			8		5
7	Fax	1	1							7	5
8	Telephone publicity	2							1	6	3
Rank	*Advertisement*										
1	Webpage		1				2	1		5	18
2	Leaflets			1		1	1			6	17
3	Stickers (Merchandise)						1			8	7
4	Radio	1								8	2
5	Newspaper ads								1	8	1
5	Newspaper articles								1	8	1
5	Articles in journals								1	8	1

Notes: The rating 7 represents a very effective measure, whereas 1 is an ineffective promotion measure. Undefined measures are rated as 1.
Source: Authors.

people directly, promotional activities within the organization and such for local people, public presentation abroad and personal letters. Amongst means of advertisement only webpages and leaflets scored high.

The results of the questionnaire study

Although all the investigated organizations manage forests and provide ES, it would be impossible to develop a single marketing strategy that applies for all of them. Every service and every project varies in size, availability of markets, and target group, and every organization, therefore, needs to develop individual marketing instruments. Although the general marketing strategy from the definition of goals to the measurement of the effects applies for all types of services, marketing instruments need to be chosen according to the individual attributes. In order to design the product, the organizations rely on their experience but hardly undertake systematic investigations of customer needs.

Image and reputation are of growing importance for individuals and businesses. The results of the questionnaire study suggest that the creation of services that produce a positive image for the buyer is an important driver for the markets of non-material ES. An example would be mining industries that compensate for damage done to the environment by supporting biodiversity conservation projects (ICMM 2005). In doing so, they improve their image, reduce the risk of getting a bad reputation and thus contribute indirectly to the financial outcome. An example of this is the Arnhem Land Fire Abatement Project being carried out in Northern Australia (see Box 8.2). Similarly, some exponents of the automobile industry buy forest areas or support plantations to compensate for the CO_2 the car fleet produces and thus improve their image. Providers of ES should consider the image factor in the marketing of their products. For example, certificates that attach such an image value to an ES may greatly increase demand for the service provision.

Box 8.2: The Arnhem Land Fire Abatement Project

This project, initiated by the Northern Territory Government of Australia aims to improve fire management on Aboriginal land in the Northern Territory whilst also addressing the problem of greenhouse gas emissions being caused by massive wildfires that occur in the territory every few years.

The practice of regular burning undertaken by Aboriginal people lessens the possibility of extreme wildfires occurring and, therefore, reduces the amount of greenhouse gases emitted. Carbon credits have been generated by these practices and are sold to various businesses as part of a voluntary carbon trading market. So far the project has been successful in giving local Aboriginal people income for their traditional fire burning practices while giving the participating companies good publicity and public profiles in supporting local indigenous people as well as addressing climate change.

Although most organizations did rate the SP technique as being important, it needs to be treated with caution as one cannot be sure whether the expressed WTP reflects the real preference of an individual for a public good, or the attitude of the answering individual towards the importance of the good or his/her other characteristics (Diamond and Hausman 1994). Furthermore, conducting these studies is time and money consuming, potentially delaying the eventual institution set-up.

The majority of Latin American organizations participating in the study sell their services directly. More than half claim to not include an intermediary. A similar study conducted by Mantau et al. (2001) found that 74.5 per cent of the 116 organizations sell their products directly. As direct distribution is an advantage in a market where there is only a small group of customers, this might be a reason for the small number of indirect trades. The market for ES is still small and personal contact is an important factor in gaining new customers. On the other hand, it seems important for small farmers and projects to encounter a powerful partner that can provide the necessary connections, reputation, and some know-how about marketing as well as the product. This picture is validated by the questionnaire study. Most respondents are either large organizations that coach smaller projects, contracting to forest owners, tenants, and small companies, or are rather small projects themselves that receive support from national or international partners.

Whereas in Europe more than 85 per cent of the participants state that they do not use personal communication (Mantau et al. 2001), Latin American projects definitively prefer it to impersonal contact. The advantage of direct communication is that the customer is obliged to react to this contact and needs to seriously consider the offer. In the case of tropical forestry projects, we see a clear trend that personal contact is the main means of promotion. Nearly all participants have established direct contact with their customers.

Of the numerous possible direct promotion methods, four seem to be more effective than the others. Direct persuasion is deemed most effective. It is followed by activities within the project and with local people. The fourth one is presenting the ideas and visions of one's organization abroad. All these are measures that demand a high personal commitment to the organization either from the organizers themselves or from well-trained staff. Large organizations seem to use a larger variety of efficiency measures than smaller ones.

Organizations with areas larger than 10,000 hectares on average, use more than six measures with a rating of seven, whereas the others on average have one. Although the number of participants is far from empirically testable, the tendency shows a clear contrast to Mantau's findings.

CONCLUSION

The research and literature on ES has, to a large extent, been one-sided. It emphasizes mechanisms that lead to the provision and supply of ES, but does not pay much attention to the creation of demand for ES. In order for PES to play a major role in improving environmental quality and reducing poverty, both supply and demand aspects of the problem must be addressed.

ES are diverse in terms of the benefits that they provide, as well as in terms of potential buyers. This suggests that a mixture of institutional arrangements and strategies is required to create demand for ES. In some cases, there is an advantage to commodifying ES and marketing them through large exchanges. In other cases, the ES are unique, and special efforts and patience for finding appropriate buyers are needed. In all cases, increasing consumer awareness of availability, value, and benefit of ES is important and leads to increased demand. Furthermore, in all cases, buyers have to be assured of product reliability, which requires explicit mechanisms for monitoring, enforcement of contracts, and insurance.

A resource manager needs to be creative in garnering PES. The diversity of ES that may be provided by individual resources should create opportunities to sell different types of ES to different buyers. This results in a high diversity of transactions (public–public, public–private, private–private) and payment mechanisms for ES (investment, credit, sales, subsidies, and donations). Many of the ES providers have mixed sources of payments and although the PES may sometimes be modest, the combined payments may allow for sustainable management of ecosystems to provide valuable environmental amenities.

REFERENCES

Alix-Garcia, A., A. de Janvry, and E. Sadoulet (2004), 'Payments for Environmental Services: To Whom, For What, and How Much?', Department of Agricultural and Resource Economics, Working Paper, University of California, Berkeley.

Babcock, Bruce A., P.G. Lakshminarayan, JunJie Wu, and David Zilberman (1996), 'The Economics of a Public Fund for Environmental Amenities: A Study of CRP Contracts', *American Journal of Agricultural Economics*, Vol. 78, No. 4, November, pp. 961–71.

Buchanan, James M., and Gordon Tullock (1975), Polluters' Profits and Political Response: Direct Controls versus Taxes', *American Economic Review*, Vol. 65, No. 1, March, pp. 139–47.

Diamond, P.A. and J.A. Hausman (1994), 'Contingent Valuation: Is Some Number better than No Number,' *The Journal of Economic Perspectives*, Vol. 8, pp. 45–64.

Feder, Gershon, Richard E. Just, and David Zilberman (1985), 'Adoption of Agricultural Innovations in Developing Countries: A Survey', *Economic Development and Cultural Change*, Vol. 33, No. 2, January, pp. 255–98.

Heiman, Amir, Bruce McWilliams, and David Zilberman (2001), Demonstrations and Money-Back Guarantees: Market Mechanisms to Reduce Uncertainty, *Journal of Business Research*, Vol. 54, pp. 71–84.

ICMM (2005), 'Biodiversity Offsets—A Briefing Paper for the Mining Industry.' International Council on Mining and Metals, London.

Kotler, Philip and Sidney J. Levy (1969), 'Broadening the Concept of Marketing', *Journal of Marketing*, Vol. 33, No. 1 January, pp. 10–15.

Mantau, U, O. Merlo, W. Sekot, and B. Welcker (2001). *Recreational and Environmental Markets for Forest Enterprises*, CABI Publishing, US.

Olmstead, Janis, David Sunding, Douglas Parker, Richard Howitt, and David Zilberman (1997), 'Water Marketing in the '90's: Entering the Electronic Age', *Choices,* Third Quarter, Vol. 26, pp. 24–8.

Rabin, M. (1998), Psychology and Economics Department of Economics, Working Paper, University of California, Berkeley, *Journal of Economic Literature*, Vol. 36, No. 1, March, pp. 11–46.

Rausser, Gordon C. (1982), 'Political Economic Markets: PERTs and PESTs in Food and Agriculture', *American Journal of Agricultural Economics*, Vol. 64, No. 5, Proceedings Issue, December, pp. 821–33.

Shostack, G.L. (1977), 'Breaking Free from Product Marketing, *Journal of Marketing*, Vol. 44, pp. 73–80.

Zusman, Pinhas (1976), 'The Incorporation and Measurement of Social Power in Economic Models', *International Economic Review*, Vol. 17, No. 2, June, pp. 447–62.

9

Valuing Improved Coastal Water Quality for Beach Recreation on the Caribbean Island of Tobago*

NESHA BEHARRY AND RICCARDO SCARPA

INTRODUCTION

The coastal and marine environment of small island states such as Tobago provides tourists with an escape from the increasing urbanization of the developed world. Beach recreation is an important contributor to welfare on this island but degradation of the environment would present a considerable threat to the dynamic and diverse population which depends on beach recreation as a source of enjoyment and income. The quality of recreational waters is a major environmental issue in Tobago, both for supporting the economically important tourism sector and for safeguarding public health.

International visitors alone contribute to over 30,000 visits to the beaches of Tobago every year.[1] The increase in visitation, combined with pollutants from land-based activities, causes coastal water pollution and degradation problems. Within the past 10 to 15 years, coastal water pollution has become an important concern, not only in Tobago but the whole Caribbean region (Siung-Chang 1997). Deterioration of coastal water quality has not only made many beaches unsuitable for swimming, but has also damaged ecological systems such as coral reefs, mangroves, and seagrass communities (IMA 2006).

The correct determination of values for public goods such as beach water quality is essential information for economically optimal environmental protection and management. In economics,

*The authors would like to thank Dr. Jim Smart of the Environment Department at the University of York for his helpful review and comments.
[1]Information on beach visitation compiled from data provided by The Central Statistical Office of Trinidad and Tobago (CSO 2002).

environmental policy decisions are commonly modelled using the assumption of homogeneity across individuals. However, while the assumption of homogeneity in individual preferences is effective for theoretical inquiries into the general properties of environmental problems, propositions based on preference homogeneity offer limited guidance on the distributional consequences of policy decisions involving national goods (Milon and Scrogin 2002).

The interactions between recreation and tourism in the marine environment are made even more complex when users have diverse preferences about marine water quality. Therefore, studies which account for this heterogeneity of taste in the measurement of preferences can be useful in making a better informed case for protection, thereby improving policy making.

Valuation studies on recreational benefits from coastal waters in the Caribbean region have largely concentrated on benefits derived from marine recreational activities such as scuba diving and snorkelling (Williams and Polunin 2000; Barker 2003; Thur 2004). While these two activities are commonly undertaken by visitors to the Caribbean, most of the local population do not regularly engage in these activities. The systematic categorization of the two recreational groups, snorkellers and non-snorkellers, was instrumental in allowing us to specifically investigate the preferences of visitors and local beach users. The objectives of this study are as follows:

1. To determine whether there are differences in the WTP values for beach access due to improved coastal water quality for two groups of beach recreationists: snorkellers and non-snorkellers, and to evaluate any differences which exist.

2. To compare two alternative ways of addressing preference heterogeneity in practice, finite mixing (latent class analysis) and continuous mixing (random parameter) of taste parameters by means of mixed logit, and

3. To evaluate policy initiatives for beach recreationists on the basis of the results of preference estimation.

RELATED LITERATURE AND CONTRIBUTION

Beach-related activity is one of the marine recreational activities[2] that has received the most attention in empirical literature. Others

---

[2]These are recreational activities that take place on the tidal estuaries, adjacent to beaches and to open waters (Freeman 1995).

include fishing, swimming, and boating (Freeman 1995). These studies are based on the premise that better water quality could improve the experience of marine recreational users, which in turn could lead to measurable economic benefits. Most of the value measures for beach recreation due to changes in water quality have been obtained by using the CVM (N. E. Bockstael and McConnell 1989; Goffe 1995), combinations of the travel cost and random utility methods (Sandstrom 1996; Hilger 2005; Soutukorvaa 2005; Parsons and Massey 2003) and, more recently, choice experiments.

Although choice experiments have become quite prevalent in environmental economics, few choice experiment studies have focussed on coastal water quality. Even fewer have examined choices about coastal water quality in the context of beach recreation. In the choice experiments undertaken to date, two main approaches have been used to examine the effect of varying levels of attributes on the beach visitor's experience. The first approach includes studies which have estimated the recreational value of beach access due to a change in a site characteristic linked to water quality. The second approach describes those with characteristics not directly linked to water quality, examples of which include congestion (McConnell 1977) and beach nourishment (Silberman 1992, Huang and Poor 2004, Landry and Keeler 2003). To date only two studies exist in the former category—(EFTEC 2002) and (Eggert and Olsson 2005). In the EFTEC (2002) study, six attributes were examined and related to the implementation of a revised European Commission Bathing Water Quality Directive. In the study by Eggert and Olsson (2005), water quality was described using four attributes. Both studies linked the levels of their attributes to changes in coastal water quality.

Investigation of heterogeneous preferences for recreational services from natural resources has been receiving increasing attention in the economic literature within the past seven years (Breffle and Morey 2000; Scarpa and Thiene 2005; Shabbar 2005). To date there have been three studies which have accounted for heterogeneous preferences in the context of beach recreation. Only one study has used choice data (Eggert and Olsson 2005), and these were analysed using a random parameter logit model. The two other studies based on travel cost data, (Hilger 2005) and (Parsons and Massey 2003), used the finite and continuous mixed logit models, respectively for the analysis. To the best of our knowledge, this research represents the third application of the choice experiment method to examine the effect of varying levels of beach visit attributes on beach users' WTP for beach access. It contributes to the literature in three ways:

1. In order to value changes to coastal water quality, it systematically categorizes beach recreationists into two groups (snorkellers and non-snorkellers) and examines their preferences separately by using different choice experiment designs.
2. It is the only choice experiment in the context of beach recreation which has used the finite mixing (latent classes) hypothesis to analyse SP data.
3. It is the first choice experiment application in the context of beach recreation that has been undertaken in a developing country.

STUDY AREA

The valuation experiments were carried out on residents and tourists at the Crown Point International Airport in Tobago. As was the case for the study done by Naidoo and Adamowicz (2005), a survey at an airport provided a sample of foreign non-nationals as well as

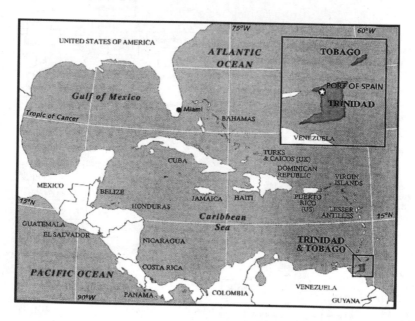

Map. 9.1:   The location of Trinidad and Tobago

Source : Wood L (2000), Lonely Planet Publications: Australia nationals who lived in Trinidad and Tobago. Tobago is located in the Caribbean, the world's most tourism-intensive region (WTTC 2004). Located in the south-east corner of the Caribbean Sea, off the coast of Venezuela, the tourism industry is very important to Tobago's economy (see Map 9.1). The Trinidad and Tobago Satellite Account for 2005 showed that the tourism industry in Tobago accounted for 46 per cent of the islands' GDP and provided 56.8 per cent of the island's employment (WTTC 2004).

Tobago is the eighth most visited island in the Caribbean and its beaches have historically been an important attraction for both overseas and domestic tourists (WTTC 2004). Tobago's attraction as a tourist destination is related to icons such as the Buccoo Reef Marine Park and Little Tobago. With 42 main beaches distributed over 300 sq. km of land, beach-related activities are the most popular in Tobago (see Map 9.2). Tourists regularly engage in snorkelling, glass bottom boat tours, and scuba diving on the 11 main coral reefs sites (Laydoo 1990) that surround the island.

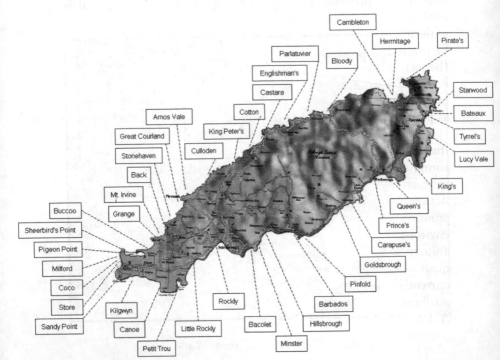

Map 9.2:   Map of Tobago showing 42 beach sites in the sample

With increase in travel, the coastal zone has experienced significant changes in recent years with respect to visitor demographics. Consequently, during the past decade, resource managers have become concerned about degradation of the reefs from local sources of pollution (Agard and Gobin 2000; Siung-Chang 1997; IMA 2006). The source of the problem has been identified as nutrient pollution leading to a series of environmental problems such as eutrophication, harmful algal blooms, loss of seagrass and coral reefs, and marine diseases (Lapointe et al. 2003). Nutrient pollution of Tobago's coastal waters has a direct impact on the tourism industry. While the environmental impacts of degraded coastal water quality have been described and documented, there has to date been no study that has estimated the recreational benefits of improving the coastal water quality on the island.

ATTRIBUTE SELECTION AND DESCRIPTION

Improvement in coastal water quality was described in terms of six attributes for the non-snorkellers and nine for the snorkellers, exclusive of the fee attribute. All attributes were described in quantitative terms to maximize the valuation potential and ease of interpretation (Bennett and Blamey 2001). Each attribute, and its level, was chosen to satisfy scientific understanding about its environmental impact on coastal water quality as also the public perception of that issue as perceived from focus groups and background work.

Three levels were used to represent each attribute according to the intensity of proposed improvement. To maintain clarity in communication, after an initial detailed description, the three levels were referred to as High Level Policy Action, Low Level Policy Action, and status quo. The high level represented the greatest amount of policy intervention which implied a high level of environmental quality while the low level represented a reduced amount of intervention and hence a lower level of environmental quality. The status quo option represented the situation as it was currently on the island. All images and wording used to portray attributes and levels were tested in several rounds of focus groups. A description of each attribute and its levels for both subsamples is presented in Table 9.1.

The underlying assumption for all attributes is that if increasing the level of the attribute contributes to improved coastal water

Table 9.1: Attribute definitions, levels, and variable names

Attribute	Definition	Variable Names and Levels	
Boats	Number of recreational and fishing boats near the coastline	BTS1_Low_Policy	Up to seven boats allowed near coastline.
		BTS2_Hgh_Policy	Up to 2 boats allowed near coastline.
Marine protected area	Presence of a marine protected area	MPA1_Low_Policy	A marine protected park where you can tour, swim, snorkel, dive, and fish.
		MPA2_High_Policy	A marine protected park where you can tour, swim, snorkel, dive but not fish.
Coastline development	Percentage of coastline development	DEV1_Low_Policy	Up to 75 per cent development allowed on the coastline.
		DEV2_High_Policy	Up to 25 per cent development allowed on the coastline.
Average bathing water quality	Percentage of coastline development	WQ1_Low_Policy	Increased chance of obtaining an ear infection from swimming in polluted water.
		WQ2_High_Policy	Reduced chance of obtaining ear infection from swimming in polluted water.
Clarity	Level of vertical visibility	CLAR1_High_Policy	Vertical visibility of up to 10 metres.
		CLAR2_Low_Policy	Vertical visibility of up to 5 metres.
Plastics	Number of plastics per 30 metres of coastline	PLAS11_Low_Policy	Up to 15 pieces per 30 metres of coastline allowed.
		PLAS2_High_Policy	Less than 5 pieces allowed per 30 metres of coastline allowed.
Snorkellers	Number of snorkellers allowed per group	SNO1_Low_Policy	Up to 15 snorkellers allowed per group or per instructor.
		SNO2_High_Policy	Up to 5 snorkellers allowed per group or per instructor.
Coral Cover	Percentage of coral cover available for viewing while snorkelling	CORAL1_High_Policy	Can view up to 45 per cent coral cover while snorkelling.
		CORAL2_Low_Policy	Can view up to 15 per cent coral cover while snorkelling.
Abundance of Fish	Number of fish available for viewing while snorkelling	FISH1_High_Policy	Can view up to 60 fishes while snorkelling.
		FISH2_Low_Policy	Can view up to 10 fishes while snorkelling.
Fee	Contribution fee to a beach authority	FEE	TT$10, TT$20, TT$25

Source: Authors.

quality, then this would increase the probability of an individual visiting the beach site. The following presents the rationale for the selection of the 10 attributes:

1. Number of Boats near the Coastline: This is the number of boats which are moored along the coastline of a beach. Boats in Tobago are used for a number of purposes, including local commercial fishing, recreational fishing, and recreational trips such as snorkelling, scuba diving, and glass bottom boat tours. With the exception of a few beaches with marinas, most of the beaches do not have launching sites which can be used by boat operators. This forces local fishermen to use the beach to store and launch their boats. As the number of boats increases near the coastline at each beach site, there is an increase in the potential for oil and fuel spills, noise (Vasconcellos and Latorre 2001), offensive odours from boat engines, and habitat disturbance such as seagrass trampling. The expectation is that a reduction in the number of boats near the coastline will have a positive impact on visitation to a beach.

2. Marine Protected Park: This is the presence of a marine park off the coastline of a beach. In this study two types of marine parks are considered, the first being one which allows fishing, and the second which does not (no take zone). The Buccoo Reef Marine Park (BRMP) is the only designated marine park on the island. However, like many marine protected areas in the Caribbean, protected area management is not enforced, nor is incentive-based management used to help in compliance with the rules of the BRMP (Mascia 1999). The expectation for this attribute is that the snorkellers' subsample will be more willing to visit the beach once a marine park is designated. This is because they would perceive its presence as improving the opportunity to view marine life. Since there is so little information on how recreationists would react to a 'no take' zone, we are not able to form a priori expectations on this attribute.

3. Development: The level of development refers to the percentage of built up environment along the coastline. The assumption here is that as development increases, the level of coastal water quality decreases and this in turn causes decreased beach visitation. This attribute was chosen because the level of development is an important indicator of the level of coastal water pollution.

4. Water Quality: Water quality is expressed through the chance of obtaining an ear infection from swimming in polluted water. This method of expression was chosen because no epidemiological

studies have been undertaken in Tobago or elsewhere in the Caribbean concerning diseases which can be obtained from swimming in polluted water. Furthermore, there is no consistent record of microbiological indicators being sampled from the beaches. Discussions with the local county medical officer indicated that ear infections are the most common ailment which he believes are linked to swimming in polluted coastal waters (Weeks 2005). The assumption here is that an increased chance of getting an infection will mean that the level of water quality is poor and will lead to a consequent decrease in visitation to the beach.

5. Clarity: This is the level of vertical visibility. The assumption here is that the level of vertical visibility increases as the quality of water increases. The expectation is that as vertical visibility increases, beach visitation will also increase.

6. Plastics: This is defined as the number of pieces of plastic found along a 30 metre stretch of coastline. Plastic is one of the most common types of marine litter and has been identified as a major issue for coastal managers (Santos et al. 2005). The expectation is that as the number of pieces of plastic increases, beach visitation will decrease.

7. Number of Snorkellers: This is the number of snorkellers allowed to snorkel per group or per instructor. Currently there are no restrictions in Tobago on the number of snorkellers per group. It is well known that the intensity of snorkelling has been linked to the coral degradation (Epstein et al. 1999; Barker 2003). Therefore, the assumption is that as the number of snorkellers increase, the probability of coral reef damage also increases with a consequent reduction in beach visitation.

8. Coral Cover: This is the amount of coral cover one can see while snorkelling. The expectation is that less coral cover will reduce snorkelling opportunities and lead to reduced beach visitation.

9. Abundance of Fish: This is the amount of fish that one can see while snorkelling. The expectation here is again that as the abundance of fish decreases, beach visitation will also decrease.

10. Fee: Because of past contentious issues associated with paying for beach access (Potts 2003), the cost attribute was described as a fee which could be paid to a non-governmental beach authority rather than the Government or private firm. This beach authority would then use the money for coastal water quality improvements. The expectation is that the implementation of a fee-based system would decrease beach visitation.

Experimental Design and Survey Administration

To evaluate the effects of preference heterogeneity, two multi-attribute stated choice survey instruments were developed, one for snorkellers and one for non-snorkellers. The development of the choice experiment questionnaires followed four stages: discussions with regional and international marine experts, focus groups, pilot studies, and final pre-testing. At each stage, the questionnaire was improved to ensure that all new information was included in the final version, and that this information could be communicated with ease to the average respondent. As the sample contained respondents from both developing and developed countries, separate focus groups were held with: (i) members of the local population (nationals); (ii) visitors to the island (non-nationals and nationals); and (iii) non-nationals who were living permanently or temporarily in Tobago.

The questionnaire comprised four subsections. The first contained questions about the frequency of use of beaches in Tobago, frequency of activities enjoyed at the beach, attitudes towards coastal water quality and management responsibility, and preferences for beach characteristics. The hypothetical scenario of the valuation section was based on respondents imagining a day out at Tobago's beaches.

Next, the attributes used in the choice experiment were explained with the aid of photographs to help respondents visualize what each attribute would look like on a beach in Tobago. Respondents were then given three options. The first two options represented beaches which they could visit with varying levels of attributes, while the third represented the base alternative. This latter alternative captured the option that they could visit another attraction or stay at home at no additional cost. Tables 9.2 and 9.3 show sample choice sets for each subsample. Respondents were presented with a worked example of a choice card. The choice sets were created using a main-effect orthogonal statistical design generated using SPSS®. Each of the two alternatives had nine attributes exclusive of cost and each attribute had three levels. We employed a main effects design consisting of 54 alternatives that were blocked into three versions with nine choice tasks, each including one pair of alternatives obtained by cycling, a technique that recent Monte Carlo studies have shown to be efficient and robust to common misspecifications (Carlsson and Martinsson 2003, Ferrini and Scarpa 2007).

Table 9.2:   Sample Choice-Set for Snorkellers

	Beach C	Beach D	Neither Beach
Boats (near coastline)	Up to 7 boats	2 Boats or less	
Marine Protected Area (absence or presence)	Yes, to visit only, no fishing	–	
Coastline Development (hotels and homes)	Up to 75% developed	Less than 25% developed	
Average Bathing Water Quality (chance of an ear infection)	–	Increased chance	
Water Clarity (down to seabed)	–	Up to 10 metres	I Choose to Visit Neither Beach
Plastics (per 30 metres of beach)	–	Up to 15 pieces	
Number of Snorkellers (per group)	–	Up to 15 per group	
Level of Coral Cover	Up to 45% coral cover	15% coral cover	
Abundance of Fish	Up to 60 fishes	Up to 10 fishes	
Contribution Fee (TT$)	TT$25.00	TT$10.00	TT$0
I would choose to visit	–	–	–

Notes: A Dash (–) represents that there is no programme to control the levels of this factor at the beach.
TT$10.00 = US$1.40 = £0.90; TT$25.00 = US$4.00 = £2.27.
Source: Authors.

The third section contained questions about why respondents were or were not willing to pay for improvements in each of the attributes. The final section gathered socio-economic characteristics such as education, income, and work status, amongst others. The sample analysed in this paper consisted of 284 completed questionnaires through face-to-face interviews, of which 198 were completed by snorkellers and 86 by non-snorkellers. Out of the snorkeller subsample, 14 per cent were nationals while 86 per cent were non-nationals. 70 per cent of the non-snorkellers were nationals while only 30 per cent were non-nationals. This varied composition of non-nationals and nationals for both subsamples highlights the fact that most nationals did not engage in snorkelling as a marine recreational activity.

Table 9.3:   Sample Choice-set for Non-snorkellers

	Beach C	Beach D	Neither Beach
Boats (near coastline)	–	Up to 7 boats	
Marine Protected Area (absence or presence)	–	Yes, to visit and fish	
Coastline Development (hotels and homes)	–	Up to 75% developed	
Average Bathing Water Quality (chance of an ear infection)	Increased chance	Reduced chance	I Choose to Visit Neither Beach
Water Clarity (down to seabed)	–	Up to 10 metres	
Plastics (per 30 metres of beach)	Less than 5 piece	–	
Contribution Fee (TT$)	TT$20.00	TT$10.00	TT$0.00
I would choose to visit	–	–	–

Notes: A Dash (–) represents that there is no programme to control the levels of this factor at the beach.
TT$10.00 = US$1.40 = £0.90; TT$25.00 = US$4.00 = £2.27.
Source: Authors.

ECONOMETRIC MODELLING AND RESULTS

Researchers have developed a number of different methods to evaluate data generated from stated choice experiments. In particular, analysis of stated choice data in environmental economics applications has benefited from recent substantial advances in discrete choice modelling. As in other fields such as transport and marketing, nearly all applications use the multinomial logit (MNL) model (McFadden 1974) which, though easy to estimate, is subject to shortcomings due to stringent assumptions and its limited ability to capture random taste heterogeneity across decision makers (McFadden and Train 2000).

The Random Parameter Logit Model

In the continuously mixed logit model, or random parameter logit (RPL) model, the error term for individual n and alternative j, ε_{nj} is assumed to be composed of two additive elements so that the utility function for alternative j as perceived by respondent n is described as:

$$U_{nj} = \beta'x_{nj} + [\eta'_{n}x_{nj} + \varepsilon_{nj}] \qquad ...(9.1)$$

The first part, η_{nj}, is an idiosyncratic random term associated with taste intensity with zero mean whose distribution over individuals and alternatives depends in general on the underlying parameters and observable data relating to alternative j and individual n. The second part, ε_{jn}, is a random term that is i.i.d. Gumbel over alternatives and does not depend on the underlying parameters or data (Brownstone and Train 1999). The underlying assumption for the RPL class of model is that η_{nj} takes a general distribution such as normal, log normal, uniform, or triangular (McFadden and Train 2000). In this study all variables were specified as normal with the exception of price which was specified as triangular.

A complete explanation of how to derive the unconditional choice probabilities for the RPL model can be found in (Train 2003) and (Hensher et al. 2005). Briefly, simulation is required in which random values are drawn from the distributions of each of the random parameters so as to integrate out the respondent-specific idiosyncratic random effects on mean taste intensities. The inclusion of a standard deviation of a beta parameter accommodates the presence of taste heterogeneity for attributes in the sampled population which does not depend on a priori information. Rather, it depends on selecting a distribution to represent the variation of respondent's preference intensities, thus presenting empirical challenges (Hensher and Greene 2003b). Further, the location of each individual's preferences is known. However, individual specific estimates can be retrieved by deriving the individual's conditional distribution based on their observed choices (Hensher and Greene 2003b; Greene et al. 2005).

The Finitely Mixed Model

As with the RPL model, this type of mixed logit model (often called latent class model, or LCM) also captures heterogeneity of taste intensities. However, in this model, heterogeneous taste intensities are employed when the researcher assumes the presence of latent variables which take the form of discrete constructs (Walker 2001). Again, we only provide an overview of the LCM, for an extensive discussion of the identification and estimation of model parameters see Heckman and Singer (1984). There are many variants of LCM, but the one used in the analysis of this study is based on the influence of choice-based attributes (as undertaken by Scarpa et al. 2005) in the estimation of latent segments as opposed to motivational factors, socio-economic covariates, and attitudinal data (as done by Boxall and Adamowicz 2002).

Briefly, the LCM is specified as a random utility model where respondent n belongs to latent class $s = (1, 2, ..., S)$. The utility function can now be expressed as:

$$V_{nj|s} = \beta_s X_{nj} + \varepsilon_{nj|s} \qquad ...(9.2)$$

Each latent class consists of a number of respondents assumed to be homogenous with respect to their preferences. By following Swait (1994), we now consider a respondent's individual segment membership likelihood function M^* which classifies respondents into one or more S segments. In this application, the classification variables influencing segment membership are only related to the latent psychographic constructs for visiting a beach. The membership likelihood function for individual n and segment s can be expressed as:

$$M^*_{ns} = \lambda_s Z_n = \varepsilon_{nj|s} \qquad ...(9.3)$$

where Z_n is a vector of psychographic constructs (P_n). See Swait (1994) for an explanation of how the parameter estimates are then derived.

Calculating WTP

The parameter estimates from the MNL, RPL, and LCM models can be used to calculate the marginal rate of substitution (or marginal willingness to pay or MWTP) for each attribute (Hanemann 1994).

$$MWTP = \frac{-\beta_k}{\beta_{cost}} \qquad ...(9.4)$$

where β_k is the coefficient on each attribute and β_{cost} is the coefficient of the price attribute.

For the LCM model, one can also derive individual-specific conditional estimates of the MWTP for each attribute k (Haab and McConnell 2002). Following the derivation in Scarpa and Thiene (2005), this can be expressed as:

$$WTP_n = \hat{E}\left(\frac{-\beta_{nk}}{\beta_{cost}}\right) = \sum_{s=1}^{S} \hat{Q}_{nk}\left(\frac{-\beta_{nr}}{\beta_{cost}}\right) \qquad ...(9.5)$$

For the RPL model, WTP measures can be constructed using either mean, conditional, or unconditional parameter estimates (Hensher et al. 2005) and using simulation to draw values from the distributions of the random parameters. In this chapter we have adopted the approach of computing these at representative values of the distribution of taste at the means.

Calculating Consumer Surplus

The consumer surplus (CS) associated with changing site characteristics measures the amount of compensation required to equate the expected utility a person receives in the altered state (state 1) in comparison to the ordinal state (state 0). The CS formula based on the MNL Model for a single choice occasion is as follows:

$$CV = \frac{\ln \sum_j \exp(\beta' x_j^1) - \exp(\beta' x_j^0)}{-\beta_{cost}} \quad ...(9.6)$$

where x^0 and x^1 are vectors of levels associated with site attributes in the original state and the altered state, respectively. For the LCM model, the CS was calculated using the individual specific conditional estimates in a similar way to that used to find the WTP in equation 9.5. For the RPL model, the values of the $\beta' x^j$ vector were replaced with the marginal rates of substitution derived from the RPL means and the MNL parameter estimates.

Estimation Results

This section describes the estimation of the parameter and WTP estimates for the MNL, LCM, and RPL models.

Model Selection and Robustness

The econometric results are presented for the continuous mixed logit (RPL) and the finitely mixed latent class logit (LCM) models for both the snorkeller and non-snorkeller subsamples. For comparison and completeness, the MNL results are also presented in Tables 9.4 and 9.5. The LCM models for both subsamples were estimated initially over 2, 3, and 4 classes. Statistical criteria, namely Akaike Information Criteria 3 (AIC-3) (Andrews et al. 2002), were used in addition to the analyst's judgement on the number of chosen segments which best described the respondent population and addressed the relevant policy questions. This analysis revealed that a two-class model provided the best solution for both the snorkeller and non-snorkeller subsamples.

The sensitivity of ML estimates to the number of draws used for simulation was explored. This analysis revealed that the model

Table 9.4: Parameter estimates from MNL, LCM, and RPL models for snorkellers

	MNL		LCM Class 1		LCM Class 2		RPL			
	Est.	t-stat	Est.	t-stat	Est.	t-stat	Est.	t-stat	Std. dev	t-stat
BTS1_Low_Policy	-0.197	-2.3	-0.147	-1.3	-0.596	-4.2	-0.330	-2.5		
BTS2_High_Policy	0.136	1.7	0.275	2.6	-0.333	-2.4	0.166	1.4		
MPA1_Low_Policy	0.125	1.5	0.369	3.4	0.221	1.5	0.291	2.1	0.180	4.141
MPA2_High_Policy	0.289	3.6	0.483	4.6	0.340	2.5	0.646	5.4		
DEV1_Low_Policy	-0.411	-4.7	-0.236	-2.2	-0.622	-4.0	-0.707	-4.5	0.185	5.155
DEV2_High_Policy	0.376	4.9	0.573	5.5	0.415	3.6	0.722	5.4	0.165	5.420
WQ1_Low_Policy	-0.421	-4.8	-0.411	-3.5	-1.259	-7.2	-0.980	-5.3	0.197	7.633
WQ2_High_Policy	0.459	5.9	0.370	3.6	0.741	6.2	0.667	5.0	0.182	4.713
CLAR1_High_Policy	0.311	3.9	0.600	5.5	0.209	1.5	0.700	5.8	0.239	2.581
CLAR2_Low_Policy	-0.215	-2.6	-0.045	-0.4	-0.281	-1.9	-0.205	-1.6	0.195	6.408
PLAS1_Low_Policy	-0.166	-1.9	0.001	0.0	-0.991	-6.4	-0.459	-2.9	0.169	5.314
PLAS2_High_Policy	0.514	6.4	0.749	6.9	0.451	3.6	0.908	6.5	0.192	5.578
SNO1_Low_Policy	-0.119	-1.4	0.070	0.7	-0.910	-5.9	-0.324	-2.2		
SNO2_High_Policy	0.108	1.3	0.221	2.1	0.139	1.1	0.118	1.0		
CORAL1_High_Policy	0.464	5.7	0.748	6.8	0.187	1.4	0.829	5.9	0.166	5.768
CORAL2_Low_Policy	-0.110	-1.3	0.113	1.1	-0.554	-3.6	-0.168	-1.3		
FISHi_High_Policy	0.240	2.9	0.447	4.2	0.147	1.1	0.556	4.5		

(contd...)

Table 9.4 (contd...)

| | MNL | | LCM | | | | RPL | | | |
| | | | Class 1 | | Class 2 | | | | | |
	Est.	t-stat	Est.	t-stat	Est.	t-stat	Est.	t-stat	Std. dev	t-stat
FISH2_Low_Policy	-0.092	-1.1	-0.008	-0.1	-0.169	-1.2	-0.082	-0.7		
Fee	-0.023	-5.8	-0.014	-2.9	-0.035	-5.5	0.058	8.0	0.007	8.043
Number of observations					5346					
Number of Individual					198					
Probability of Membership			61%		39%					
Number of Parameters (K)	19		39		30					
Log Likelihood (LL):	-1742.915		-1578.906	-1604.02						
Akaike Information—3	5285.75		4853.72	4902.06						

Source: Authors.

Table 9.5: Parameter estimates from MNL, LCM, and RPL models for non-snorkellers

	MNL		LCM				RPL			
			Class 1		Class 2					
	Est.	t-stat	Est.	t-stat	Est.	t-stat	Est.	t-stat	Std. dev	t-stat
BTS1_Low_Policy	0.179	1.5	0.346	2.6	-2.853	-3.4	0.5227	3.0870		
BTS2_High_Policy	0.397	3.0	0.488	3.5	0.804	1.1	0.8670	4.9960		
MPA1_Low_Policy	-0.025	-0.2	0.128	1.0	-34.050	0.0	0.0118	0.0590	0.927	3.8
MPA2_High_Policy	0.238	1.9	0.377	2.8	2.763	0.0	0.6708	3.9230		
DEV1_Low_Policy	-0.197	-1.5	-0.114	-0.8	-3.585	-3.1	-0.3873	-1.4950	1.501	5.3
DEV2_High_Policy	0.274	2.2	0.339	2.4	0.907	1.3	0.6729	4.0130		
WQ1_Low_Policy	-0.622	-4.6	-0.575	-4.0	-6.177	0.0	-1.0880	-4.6920	1.118	4.3
WQ2_High_Policy	0.487	4.0	0.580	4.3	31.395	0.0	0.9405	4.7820	0.951	4.0
CLAR1_High_Policy	0.469	3.9	0.602	4.7	0.775	1.0	0.7247	4.2680		
CLAR2_Low_Policy	0.031	0.2	0.114	0.8	-0.015	0.0	0.1855	0.9320	0.631	2.2
PLAS1_Low_Policy	-0.312	-2.4	-0.201	-1.5	-34.062	0.0	-0.4215	-2.5240		
PLAS2_High_Policy	0.296	2.4	0.414	3.0	-33.146	0.0	0.5874	3.4610		
Fee	-0.025	-3.7	-0.018	-2.6	-0.018	-2.6	0.0942	6.6270	0.094	6.6
Number of observations					2193					
Number of Individuals					86					
Probability of Membership			83%	17%						
Number of Parameters (K)	13		27				19			
Log Likelihood (LL):	-747.35		-660.7526	-684.631						
Akaike Information—3	2281.05		2063.26	2110.89						

Source: Authors.

based on 300 draws provided sufficiently good approximations for the estimates from both subsamples. In both models, all attributes were first specified as random using the normal distribution. In order to ensure non-negative parameter estimates, the fee parameter was specified as log normal and all other attributes for both models were specified as normal. The results from these estimations revealed a number of parameters with insignificant standard deviations (SD). This was used as the basis for selecting the random parameters (Hensher et al. 2005). Derivation of the WTP estimates for both restricted models specifying the fee parameter as log normal yielded implausible WTP estimates. The fee parameter was then specified using the constrained triangular distribution which led to more behaviourally plausible WTP estimates and also achieved the goal of a sign-constrained cost parameter (Hensher and Greene 2003b). Therefore, the final model was estimated with only the attributes which had significant SD. These were all specified as normal with the exception of the fee parameter which was specified as random with a constrained triangular distribution.

Snorkeller Subsample—Parameter Estimates and Preference Groups

The LCM Model is estimated with 2, 3, and 4 segments. The 4 segment model did not converge. Based on the AIC-3 test statistic in Table 9.6, the 3 segment model had the greatest explanatory power but class two of the 3 segment model had a positive fee parameter. Therefore, to be consistent with economic theory, the 2 segment model was chosen because both classes had negative estimates for the fee parameter. Table 9.4 shows the estimation results for this model.

The estimation results for the 2 segment model showed that there were dramatic differences between the two classes. Individuals in class 1 are likely to choose beaches which allow up to 2 boats, provide access to an marine-protected park (whether it allows fishing or not), have clarity of 10 m, allow up to 5 persons per snorkelling group, provide up to 45 per cent coral cover and the ability to view up to 60 fishes while snorkelling. Individuals with group 2 preferences are likely to choose beaches which do not allow any boats, have a marine-protected park which allows fishing, allow up to 15 persons per snorkelling group and provide up to 15 per cent coral cover.

The most positive and highly significant parameter estimate for group 1 indicates that individuals in this group have the strongest preference for beaches with less than 5 plastics per 30 metres of

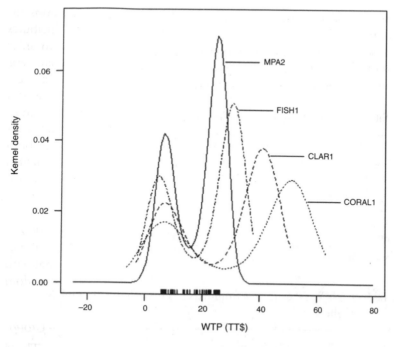

Fig. 9.1:   Individual Specific Conditional Estimates for the LCM Model

coastline. For group 2 individuals, the strongest preference is for beaches with a reduced chance of getting an ear infection. Both groups 1 and 2 had six parameters which were statistically insignificant.

Snorkeller Subsample—Distribution of Posterior WTP Estimates from the LCM Model

The conditional WTP estimates were calculated for each attribute using the parameter estimates from the LCM Model. Summary statistics are provided for each attribute in Table 9.8. In Figure 9.1, four attributes are depicted using a kernel density graph to illustrate some of the drastic differences in WTP values between the two groups. All attributes exhibit this bi-modal distribution as a consequence of the sample being composed of respondents belonging to two classes with different taste intensities.

The results from the LCM model reveal heterogeneity within the snorkeller subsample. Therefore, this subsample can be further classified into two groups. The first group representing 61 per cent of the sampled population seems to be composed of more avid snorkellers because of their strong preferences and higher WTP for high levels of fish, coral cover, water clarity, and both types of marine-protected parks. The second group, representing 39 per cent of the population, could be classified as the more occasional snorkellers with individuals who do not have very strong preferences for the presence of coral,

Table 9.6:   Model specification for snorkellers

Parameter	MNL	LCM (2)	LCM (3)	ML
Log Likelihood	−1742.92	−1578.91	−1542.25	−1604.02
Number of Parameters (K)	19	39	59	30
Number of Individuals (N)	198	198	198	198
AIC	3523.83	3235.81	3202.50	3268.04
AIC-3	5285.75	4853.72	4803.75	4902.06
BIC	3586.31	3364.05	3396.51	3366.69

Source: Authors.

Table 9.7:   WTP estimates from MNL, LCM, and RPL models for snorkellers

| | MNL | | LCM | | | | RPL | |
| | | | Class 1 | | Class 2 | | | |
	Est.	t-stat	Est.	t-stat	Est.	t-stat	Est.	t-stat
BTS1_Low_Policy	−8.711	−2.1	−10.219	−1.2	−17.152	−3.2	−5.655	−2.4
BTS2_High_Policy	6.031	1.7	19.058	2.1	−9.585	−2.0	2.839	1.4
MPA1_Low_Policy	5.553	1.5	25.605	2.4	6.355	1.5	4.975	2.1
MPA2_High_Policy	12.807	3.3	33.491	2.7	9.782	2.4	11.062	5.0
DEV1_Low_Policy	−18.196	−3.5	−16.409	−1.7	−17.903	−3.2	−12.103	−3.9
DEV2_High_Policy	16.653	4.1	39.763	2.8	11.946	3.1	12.363	5.0
WQ1_Low_Policy	−18.643	−3.6	−28.536	−2.3	−36.206	−4.1	−16.785	−4.5
WQ2_High_Policy	20.327	4.4	25.705	2.4	21.320	4.7	11.425	4.6
CLAR1_High_Policy	13.787	3.5	41.613	2.8	6.011	1.5	11.990	5.4
CLAR2_Low_Policy	−9.502	−2.3	−3.157	−0.4	−8.084	−1.8	−3.504	−1.5
PLAS1_Low_Policy	−7.335	−1.8	0.043	0.0	−28.509	−3.8	−7.858	−2.7
PLAS2_High_Policy	22.744	4.6	51.957	2.8	12.966	3.2	15.553	5.8
SNO1_Low_Policy	−5.275	−1.4	4.867	0.7	−26.177	−3.7	−5.546	−2.1
SNO2_High_Policy	−0.006	−1.3	−0.012	−2.1	4.010	1.0	−5.546	−2.1
CORAL1_High_Policy	20.571	4.5	51.876	2.9	5.372	1.4	14.200	5.3
CORAL2_Low_Policy	−4.869	−1.2	7.850	1.1	−15.937	−2.9	−2.876	−1.3
FISH1_High_Policy	10.636	2.8	30.985	2.6	4.229	1.1	9.520	4.3
FISH2_Low_Policy	−4.054	−1.0	−0.555	−0.1	−4.876	−1.2	−1.397	−0.7

Source: Authors.

Table 9.8:   Summary statistics for individual specific conditional WTP estimates from LCM model for snorkellers and non-snorkellers

	Snorkellers			Non-snorkellers		
	Average	Median	SD	Average	Median	SD
BTS1_Low_Policy	–12.897	–10.571	3.095	15.473	18.705	7.077
BTS2_High_Policy	7.996	17.606	12.787	21.836	26.398	9.988
MPA1_Low_Policy	18.171	24.629	8.594	5.725	6.921	2.619
MPA2_High_Policy	24.334	32.289	10.585	16.888	20.416	7.724
DEV1_Low_Policy	–16.986	–16.485	0.667	–5.086	–6.149	2.327
DEV2_High_Policy	29.020	38.353	12.418	15.179	18.350	6.943
WQ1_Low_Policy	–31.498	–28.925	3.424	–25.729	–31.104	11.768
WQ2_High_Policy	24.011	25.482	1.957	25.952	31.374	11.871
CLAR1_High_Policy	27.864	39.808	15.894	26.970	32.604	12.336
CLAR2_Low_Policy	–5.060	–3.407	2.199	5.118	6.188	2.341
PLAS1_Low_Policy	–10.984	–1.405	12.747	–9.020	–10.905	4.126
PLAS2_High_Policy	36.899	49.980	17.407	18.513	22.381	8.468
SNO1_Low_Policy	–7.122	3.292	13.859			
SNO2_High_Policy	1.542	0.192	1.795			
CORAL1_High_Policy	33.916	49.518	20.761			
CORAL2_Low_Policy	–1.337	6.643	10.619			
FISH1_High_Policy	20.652	29.628	11.945			
FISH2_Low_Policy	–2.224	–0.774	1.929			

Source: Authors.

fish, and clarity attributes. Both groups, however, have strong preferences for high water quality, low development, and little littering on the beach. It can be seen from Table 9.8 that snorkellers are willing to pay the most for a beach which has less than 5 pieces of plastic and are willing to pay the most for a decreased chance of infection from swimming in polluted water. Socio-economic covariates and attitudinal variables could be used to further understand heterogeneity.

Snorkeller Subsample—Preferences and WTP Estimates for the RPL Model

Table 9.4 shows the parameter estimates of the RPL model for the snorkeller subsample. All estimated coefficients are found to be of the expected sign. Again, the RPL model revealed that snorkellers are willing to pay to visit a beach which has both types of marine-protected parks. 11 of the 19 parameters provide evidence of significant taste heterogeneity by producing statistically significant SD. The attribute in which preferences vary the most for the snorkeller subsample is the increased chance of getting an ear infection. The fee attribute represents the parameter in which preferences vary the least. Examining the WTP estimates calculated at the mean of the taste distribution, we can see that snorkellers have the highest WTP estimate for beaches with up to 5 pieces of plastic, closely followed by beaches with 45 per cent coral cover. These results are presented in Table 9.7.

Non-snorkeller Subsample—Parameter Estimates and Preference Groups

The LCM was estimated for the non-snorkellers over 2, 3, and 4 segments. The four segment model did not converge. The 3 segment model had one class where the fee parameter was theoretically implausible because of a positive fee parameter. In the 2 segment model, even though the fee parameters were both negative, the fee parameter was insignificant in class 1. In order to be consistent with economic theory, the fee parameter was constrained to be equal to that of class 2 and these two groups were then analysed. Table 9.5 shows the estimation results for this model and Table 9.9 shows the AIC-3 statistical criteria.

In the 2 segment model, individuals in class 1 prefer beaches that have water clarity of 10 metres, a low chance of an ear infection, up to 2 boats, up to 5 pieces of plastic, up to 7 boats, and up to 25 per cent development. The preferences in class 2 were not very clear as 9 out of the 12 parameter estimates are insignificant. However, for the parameters that were significant, it can be seen that individuals in group 2 will not visit beaches which have up to 7 boats and up to 25 per cent development.

Group 1 represents 83 per cent of the sampled population while group 2 represents 17 per cent of the population. The results indicate that by constraining the fee parameter and imposing 2 segments, the majority of people fit into group 1 while the remainder is too small to really produce significant estimates. This indicates that the non-snorkeller subsample is a relatively homogenous one when compared to the snorkeller subsample.

Non-Snorkeller Subsample—Distribution of Posterior WTP Estimates from LCM Model

The conditional WTP estimates were also calculated for each attribute for the non-snorkeller subsample using the parameter estimates from the LCM Model. Summary statistics are provided for each attribute in Table 9.8. This shows that persons were willing to pay the most for beaches with water clarity of up to 10 metres.

Non-Snorkeller Subsample—Preferences and WTP Estimates for RPL Model

Table 9.5 shows the results of the RPL Model for the non-snorkeller subsample. While the LCM model suggested that there was little preference heterogeneity within the non-snorkeller population, the RPL model reveals a different scenario. Six of the 13 parameters had significant SD. This indicates that preferences do in fact vary over this population for the following attributes: a marine park which allows fishing, up to 75 per cent development, increased and decreased chance of getting an ear infection, clarity of 5 metres, and the fee.

The attribute in which preferences vary the most for the non-snorkeller subsample is that representing up to 75 per cent development near the beach. This attribute represents a low level of policy development and could be due to the fact that most of the non-snorkeller population was made up of locals, some of which probably perceive development as a good attribute rather than as a bad one. There were three mean parameter estimates which were

Table 9.9:   Model specification for non-snorkellers

Parameter	MNL	LCM (Restricted)	LCM (2)	LCM (3)	ML
Log Likelihood	−747.35	−660.75	−660.18	−631.77	−684.631
Number of Parameters (K)	13	27	27	41	19
Number of Individuals (N)	86	86	86	86	86
AIC	1520.70	1375.51	1374.37	1345.54	1407.26
AIC-3	2281.05	2063.26	2061.55	2018.31	2110.89
BIC	1552.61	1441.77	1440.63	1446.17	1453.89

Source: Authors.

Table 9.10:   WTP Estimates from MNL, LCM, and RPL Models for Non-snorkellers

| | MNL | | LCM | | | | RPL | |
| | | | Class 1 | | Class 2 | | | |
	Est.	t-stat	Est.	t-stat	Est.	t-stat	Est.	t-stat
BTS1_Low_Policy	7.118	1.4	18.705	2.0	-154.401	-2.0	5.547	3.0
BTS2_High_Policy	15.763	2.9	26.398	2.6	43.541	1.1	9.200	4.9
MPA1_Low_Policy	-1.008	-0.2	6.921	0.9	-1842.923	0.0	0.125	0.1
MPA2_High_Policy	9.458	1.9	20.416	2.2	149.551	0.0	7.118	4.0
DEV1_Low_Policy	-7.832	-1.4	-6.149	-0.7	-194.011	-1.9	-4.110	-1.4
DEV2_High_Policy	10.887	2.3	18.349	2.3	49.068	1.1	7.141	4.3
WQ1_Low_Policy	-24.704	-3.0	-31.104	-2.3	-334.331	0.0	-11.545	-3.8
WQ2_High_Policy	19.362	3.3	31.374	2.6	1699.212	0.0	9.980	4.7
CLAR1_High_Policy	18.633	2.9	32.604	2.4	41.943	1.0	7.689	3.9
CLAR2_Low_Policy	1.249	0.2	6.187	0.8	-0.820	0.0	1.968	1.0
PLAS1_Low_Policy	-12.384	-2.0	-10.905	-1.3	-1843.590	0.0	-4.473	-2.3
PLAS2_High_Policy	11.762	2.3	22.380	2.3	-1793.985	0.0	6.234	3.5

Source: Authors.

insignificant: MPA1, DEV1, and CLAR2. Table 9.10 gives WTP values which were calculated at the mean of the estimated distributions. These values reveal that the non-snorkellers had the most positive WTP for a decreased chance of infection and the most for high vertical visibility.

Welfare Estimates for Beach Improvements

To evaluate respondents' valuation of a potential improvement scenario for the beaches in Tobago, the estimated MNL, LCM, and RPL models were used to compute CS estimates for both subsamples. The potential improvement scenario constructed involved the maximum improvement possible at all 42 beach sites and for all attributes for both subsamples. This involved independently changing each of the beach attributes from its base level condition to the highest level attainable for that attribute as defined in this experiment. Each beach had to be described and coded in terms of the attribute descriptors used in this experiment to represent the current status of what an individual would experience if they visited a beach.

The highest level of improvement possible for each attribute varied on the beach site and had to be coded and described separately as well. For example, maximum improvement of the 'Store Beach' site for the non-snorkeller subsample for the number of boats attribute involved changing the base level from the status quo scenario to one where the high level of policy could be experienced. In the case of boats, the maximum improvement would be having up to two boats near the coastline.

The CS values for both subsamples are presented in Table 9.11. For the snorkeller subsample in the MNL model, respondents were willing to pay TT98.71[3] while respondents of the non-snorkeller population were willing to pay TT$68.67. For the snorkeller subsample respondents in the LCM model, the value was calculated based on the individual specific conditional WTP estimates. This yielded a CS estimate of TT$150.85. For the non-snorkeller subsample, as observed in the MNL results, the value is also lower than that of the snorkellers at TT$91.66. The distribution for the snorkellers and non-snorkellers for this improvement policy scenario

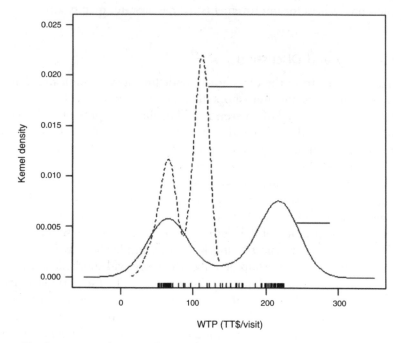

Fig. 9.2:   Distribution of WTP for Maximum Improvement Scenario

[3]The currency used is Trinidad and Tobago Dollars (TT$). Approximately 11.9 TT dollars are equal to GB£ 1.00 or US$6.25

Table 9.11: Cost per choice occasion

	MNL	LCM			RPL
		Mean	Median	Std Dev	
Snorkellers	98.71	150.86	200.39	73.84	36.42
Non-snorkellers	68.67	91.67	106.41	25.56	66.58

Source: Authors.

derived from the individual specific conditional WTP estimates from the LCM model are shown in Figure 9.2. The CS using the estimates from the RPL model were calculated at the means. This yielded a value of TT66.58 for the snorkeller subsample and TT$36.42 for the non-snorkeller.

The results reported for the CS estimates provide strong evidence that the classification of beach visitors based on their involvement in the recreational sport of snorkelling has a significant influence on their preferences for improvements to coastal water quality attributes at a beach.

Summary and Discussion

The evidence from the two choice modelling surveys indicates that both snorkellers and non-snorkellers prefer to visit beaches where there is a higher level of environmental quality as opposed to a lower one. The use of the LCM and RPL models in the analysis allowed us to further examine the preferences of the two groups of recreationists.

For the LCM, we estimated a model analogous to the mixed logit model. However, instead of representing unobserved heterogeneity with a continuous mixture of parameters, we specified two distinct groups of individuals, each group with its own set of parameters for the 6 and 9 attributes of the alternatives depending on the subsample. The LCM suggests that the snorkeller subsample comprises at least two groups. The first group describes the more avid snorkellers while the second describes the more occasional ones. The use of two latent classes does provide a significant improvement in the fit over the MNL model and the RPL model. This can be explained by the fact that there seems to be a behavioural justification for different types of recreationists, even within the snorkeller group.

For the non-snorkeller subsample, there is a similar, although rather weaker suggestion of the existence of two classes. The use of the two latent classes does provide a significant improvement in the fit over the MNL model, but the level of fit is below that captured by the RPL model. However, looking closely at the results of the LCM model, it is clear that most of the non-snorkellers fall within one class. This is not surprising as we did not have any strong behavioural justification for two distinct segments of the non-snorkellers. Therefore, in this situation, a continuous distribution provides more explanatory power.

This study has used both LCM and RPL models to understand unobserved heterogeneity for the subsamples. These two models offer alternative ways of modelling heterogeneity. Since there is no unambiguous way to determine whether one model is superior to another (Hensher and Greene 2003a), we should not focus on this issue. Rather, the policy maker should make full use of the information which both models provide. For example, in the case of the snorkellers' subsample, the results of the LCM provided information that, for this group of recreationists, the use of recreational group policies might be best suited. Alternatively, as in the case of non-snorkellers, if preferences do not vary within defined segments, then policy makers might want to use the information given by the degree of spread around the mean of each attribute. This will allow them to understand and hence focus on a specific attribute for that population.

The generation of these estimates provides a starting point for policy makers in Tobago. Three important points emerge for policy makers from these results. The first is that across the two recreational groups, both have positive preferences for specific attributes. These are the attributes that indicate a low amount of development, a reduced chance of infection from swimming in coastal waters, and a reduced level of plastics on the beach. Policy makers could concentrate their improvement efforts on the attributes which appeal to both groups first, before turning their attention to policies which deliver value to subgroups of recreationists.

The second point which emerges from this study is that group-specific recreation management policies could be designed. In this study, the LCM values showed that one group had higher WTP values and expressed stronger positive preferences for certain attributes than the other. Therefore, the introduction of restricted access at certain sites based on a pricing policy could be one way to satisfy the demand for this type of access.

Prior to this study, overviews of the preferences of beach visitors for coastal water quality in Tobago were based on perception and anecdotal evidence. Indeed, a third important finding of this study is that it indicates that respondents are less inclined to visit beaches which exhibit levels of attributes associated with low levels of policy (albeit this being in a hypothetical situation). This suggests that beach recreationists would readily seek alternative locations for recreation if the deterioration of water quality continues. This information should provide an incentive for policy makers to focus their policy planning accordingly.

REFERENCES

Agard, J.B.R. and J. Gobin (2000), 'The Lesser Antilles, Trinidad and Tobago', *Seas at the Millenium: An Environmental Evaluation*, pp. 591–605.

Andrews, R., A. Ansari, and I. Currim (2002), 'Hierchical Bayes vs. Finite Mixture Conjoint Analysis Models: A Comparison of Fit, Prediction, and Partworth Recovery', *Journal of Marketing Research*, Vol. 39, No. 1, pp. 87–98.

Barker, N.H.L. (2003), 'Ecological and Socio-Economic Impacts of Dive and Snorkel Tourism in St. Lucia, West Indies', PhD thesis, University of York.

Bennett, J. and R. Blamey (eds) (2001), *The Choice Modelling Approach To Environmental Valuation*, Edward Elgar Publishers, UK.

Boxall, P.C. and W.L. Adamowicz, (2002), 'Understanding Heterogeneous Preferences in Random Utility Models: A Latent Class Approach', *Environmental and Resource Economics 2002*, Vol. 23, No. 4, 421–46.

Bockstael, N.E. and K.E. McConnell, I.E.S. (1989), 'Measuring the Benefits of Improvements in Water Quality: The Chesapeake Bay', *Marine Resource Economics*, Vol. 6, pp. 1–18.

Breffle, W. and E. Morey (2000), 'Investigating Preference Heterogeneity in a Repeated Discrete Choice Recreation Demand Model of Atlantic Salmon Fishing', *Marine Resource Economics*, Vol. 15, pp. 120.

Brownstone, D. and K. Train (1999), 'Forecasting New Product Penetration with Flexible Substitution Patterns', *Journal of Econometrics*, Vol. 89, pp. 109–29.

Carlsson, F. and P. Martinsson (2003), 'Design Techniques for Stated Preference Methods in Health Economics', *Health Economics*, Vol. 12, pp. 281–94.

CSO (2002), *Statistics on Visitor Arrivals by Purpose of Visit in 2002*, The Central Statistical Office of Trinidad and Tobago (CSO).

EFTEC (2002), *Valuation of Benefits to England and Wales of a Revised Bathing Water Quality Directive and Other Beach Characteristics Using the Choice Experiment Methodology*, Technical Report, Economics for the Environment Consultancy Limited, Department for Environment, Food and Rural Affairs.

Eggert, H. and B. Olsson (2005), 'Heterogeneous Preferences for Marine Amenities: A Choice Experiment Applied to Water Quality', Department of Economics, Goteborg University, Working Paper.

Epstein, N., R. Back, and B. Rinkevich (1999), 'Implementation of a Small Scale "no-use zone" Policy in a Reef Ecosystem: Eilat's Reef-lagoon Six Years Later', *Coral Reefs*, Vol. 18, pp. 327–32.

Ferrini, S. and R. Scarpa (2007), 'Designs With A-Priori Information for Non-market Valuation with Choice-Experiments: A Monte Carlo Study', *Journal of Environmental Economics and Management*, Vol. 53, pp. 342–63.

Freeman, A.M. (1995), 'The Benefits of Water Quality Improvements for Marine Recreation: A Review of the Empirical Evidence', *Marine Resource Economics*, Vol. 10, pp. 385–406.

Goffe, P.L. (1995), 'The Benefits of Improvements in Coastal Water Quality: A Contingent Approach', *Journal of Environmental Management*, Vol. 45, pp. 305–17.

Greene, W., D.A. Hensher, and J. Rose (2005), 'Using Classical Simulation-Based Estimators to Estimate Individual WTP Values', *Applications of Simulation Methods in Environmental and Resource Economics*, Chapter 2, pp. 17–32.

Haab, T.C. and K.E. McConnell (2002), *Valuing Environmental and Natural Resources*, Edward Elgar Publishers, UK.

Hanemann, W. M. (1994), 'Valuing the Environment Through Contingent Valuation', *The Journal of Economic Perspectives*, Vol. 8, pp. 19–43.

Heckman, J. and B. Singer (1984), 'Econometric Duration Analysis', *Journal of Econometrics*, Vol. 24, pp. 63–32.

Hensher, D.A., J.M. Rose, and W.H. Greene, (2005), *Applied Choice Analysis—A Primer*, Cambridge University Press.

Hensher, D. and W. Greene (2003a), 'A Latent Class Model for Discrete Choice Analysis: Contrasts with Mixed Logit', *Journal of Transportation Research*, Vol. 37.

_____ (2003b), 'The Mixed Logit Model: The State of Practice', *Journal of Transportation*, Vol. 30, pp. 133–76.

Hilger, J. (2005), 'The Impact of Water Quality on Southern California Beach Recreation: A Finite Mixture Model Approach', Working Paper, University of California, Berkeley.

Huang, J.-C. and P. Poor (2004), 'Welfare Measurement with Individual Heterogeneity: Economic Valuation of Beach Erosion Control Programmes', Working Paper, University of New Hampshire.

IMA (2006), 'Background Paper by The Institute of Marine Affairs,

Trinidad and Tobago (IMA): Development of a National Programme of Action (NPA) for the Protection of the Marine Environment from Land-based Sources and Activities'.

Landry, C.E., A. Keeler, W.K. (2003), 'An Economic Evaluation of Beach Erosion Management Alternatives', *Marine Resource Economics*, Vol. 18, pp. 105–27.

Lapointe, B. (2003), 'Impacts of Land-Based Nutrient Pollution on Coral Reefs of Tobago', Report Prepared for the Buccoo Reed Trust of Tobago.

Laydoo, R.S. (1990), 'The Shallow Water Scleractineans (stony corals) of Tobago, West Indies', *Journal of Caribbean Marine Studies*, Vol. 1, pp. 29–36.

Mascia, M.B. (1999), 'Governance of Marine Protected Areas in the Wider Caribbean: Preliminary Results of an International Mail Survey', *Coastal Management*, Vol. 27, pp. 391–402.

McConnell, K. (1977), 'Congestion and Willingnes to Pay Use', *Land Economics*, Vol. 53, pp. 185–95.

McFadden, D. (1974), 'Conditional Logit Analysis of Qualitative Choice Behavior', *Frontiers of Econometrics*, Academic Press.

McFadden, D. and K. Train, (2000), 'Mixed MNL Models for Discrete Response', *Journal of Applied Econometrics*, Vol. 15, pp. 447–70.

Milon, J. and D. Scrogin (2002), 'Heterogeneous preferences and Complex Environmental Goods: The Case of Ecosystem Restoration', *Recent Advances in Environmental Economics*, Edward Elgar, Chapter 13, pp. 286–308.

Naidoo, R. and W. Adamowicz (2005), 'Biodiversity and Nature-Based Tourism at Forest Reserves in Uganda', *Environment and Development Economics*, Vol. 10, pp. 159–78.

Parsons, G.R. and D.M. Massey (2003), *A Random Utility Model of Beach Recreation*, The New Economics of Outdoor Recreation, Edward Elgar, Chapter 12, pp. 241–67.

Potts, A. (2003), 'Conflict Resolution Between Inter-Sectoral Stakeholders for the Buccoo Reef Marine Park Coastal Zone in Tobago: The Pigeon Point Case Study,' Unpublished PhD Thesis.

Sandstrom, M. (1996), Recreational Benefits from Improved Water Quality: A Random Utility Model of Swedish Seaside Recreation, Working Paper No. 121, Stockholm School of Economics, The Economics Research Institute.

Santos, I.R., A.C. Friedrich, M. Wallner-Kersanach, and G. Fillmann (2005), 'Influence of Socio-Economic Characteristics of Beach Users on Litter Generation', *Ocean and Coastal Management*, Vol. 48, pp. 742–52.

Scarpa, R. and M. Thiene (2005), 'Destination Choice Models for Rock Climbing in the Northeastern Alps: A Latent-class Approach Based on Intensity of Preference', *Land Economics*, Vol. 85, No. 3, pp. 426–44.

Scarpa, R., K.G. Willis, and M. Acutt (2005), 'Individual-specific Welfare Measures for Public Goods: A Latent Class Approach to Residential Customers of Yorkshire Water,' in P. Koundouri (ed.), *Econometrics Informing Natural Resource Management*, Edward Elgar Publishers, UK.

Shabbar, A.A.J. (2005), 'Heterogeneous Preferences for Greek Heritage Attractions', *Tourism Economics*, pp. 225–45.

Silberman, J. and D.A. Gerlowski, N.A.W. (1992), 'Congestion and Willingnes to Pay Use', *Land Economics*, Vol. 68, pp. 225–36.

Siung-Chang, A. (1997), 'A Review of Marine Pollution Issues in the Caribbean', *Environmental Geochemistry and Health*, Vol. 19, pp. 44–5.

Soutukorvaa, A. (2005), 'The Value of Improved Water Quality: A Random Utility Model of Recreation in the Stockholm Archipelago', Working Paper for The Beijer International Institute of Ecological Economics, The Royal Swedish Academy of Sciences.

Swait, J. (1994), 'A Structural Equation Model of Latent Segmentation and Product Choice for Crosssectional Revealed Preference Data', *Journal of Retailing and Consumer Service*, Vol. 1, pp. 77–89.

Thur, S.M. (2004), 'Valuing Recreational Benefits in Coral Reef Marine Protected Areas: An Application to the Bonaire National Marine Park,' PhD thesis, University of Delaware.

Train, K. (2003), *Discrete Choice Methods with Simulation*, Cambridge University Press, New York.

Vasconcellos, J. and R. Latorre (2001), 'Recreational Boat Noise Level Evaluation', *Ocean Engineering*, Vol. 28, pp. 1309–24.

Walker, J.L. (2001), 'Extended Discrete Choice Models: Integrated Framework, Flexible Error Structures, and Latent Variables,' PhD thesis, Massachusetts Institute of Technology.

Weeks, M. (2005), 'Personal communication'.

Williams, I.D. and N.V. Polunin (2000), 'Differences Between Protected and Unprotected Reefs of the Western Caribbean in Attributes Preferred by Dive Tourists', *Environmental Conservation*, Vol. 27, No. 4, pp. 382–91.

WTTC (2004), *The Impact of Travel and Tourism on Jobs and the Economy*, Technical Report, Trinidad and Tobago, World Travel and Tourism Council.

10

Analysis of Farmers' Willingness to Pay for Agrobiodiversity Conservation in Nepal

DIWAKAR POUDEL AND FRED H. JOHNSEN

INTRODUCTION

Nepal has a rich diversity of crop genetic resources (CGRs); it contains more than 2000 rice landraces including wild relatives (Gupta et al. 1996; Gauchan 1999). Conservation of such CGRs has high priority on the conservationists' agenda. The Convention on Biological Diversity (CBD) has also emphasized the need for conservation of such genetic resources (UNEP 1994). Although priority is accorded to conservation ex situ at the national level, in situ conservation of such genetic diversity has been initiated to strengthen the so-called 'de facto conservation' through national and international agencies (Jarvis et al. 2000). The de facto conservation is governed by the apparent attributes of the CGRs. The important attributes of CGRs are production, consumption, ecological adaptability, and social, cultural, and economical diversity (Brush 2000). However, how long the de facto conservation will be continued is uncertain due to land conversion and invasion of exotic species (Turpie 2003). The present importance (the use value), and the existence or future values (the non-use values) of genetic diversity are among the parameters affecting the de facto conservation of CGRs in situ. Considering this, many efforts have been made to conserve these genetic resources either in situ or ex situ. The

This paper is an output of a Master of Science thesis research. The study was conducted under a research grant provided by the Norwegian Agency for International Development and Cooperation (NORAD), Norway. The authors would like to express their sincere gratitude to the organization. The authors are also grateful to the workshop organizing committee for accepting this paper.

conservation of biodiversity particularly CGRs is possible though different approaches such as ex situ, in situ (household level), and community gene bank (see Worede 1992; UNEP 1995; Rijal 2002; Hammer et al. 2003; Rana et al. 2003; Sthapit et al. 2003; Sthapit and Jarvis 2003). The conservation strategy could be designed based on the people's preferences for the different conservation methods and the value that they assign to the genetic resources.

The value that individuals and societies place on conservation is measured by their willingness to forgo the benefits of alternative uses of the same resources. This is reflected by a measure of their WTP to acquire benefits or satisfaction from the resources (UNEP 1995). Measuring the people's WTP for genetic resources is difficult because no market exists for such resources. Economists generally choose one of two methods, either to use a surrogate market, the RPs, such as travel cost method (TCM) or to use the simulated market, the SPs such as CVM to value 'public goods' such as biodiversity (UNEP 1995). The reasons for using CVM are non-existence of surrogate market and the opinion that non-use values have to be valuated, which may elicit these non-use values (Nunes and Van den berg 2001; OECD 2002 Bräuer 2003). The CVM tries to solve this problem by creating a hypothetical market in which consumers can express their preferences in monetary terms (Mitchell and Carson 1989; UNEP 1995). By deriving WTP values for different conservation approaches and types of resources, these can be compared to provide information to policy makers on the relative values of the genetic resources to the societies (White et al. 2001).

The study uses the CVM to assess the WTP for conservation of CGRs, particularly rice landraces in three different conservation approaches. The study investigates and analyses how the people's WTP differs with the importance of the landraces, the perceived threat of extinction of the landraces, and different methods of conservation. The main purpose is to provide a basis for designing conservation strategies for CGRs.

MATERIALS AND METHODS

The CVM

The CVM is a method of valuing environmental goods for which no market prices exist. This method involves setting up a hypothetical market, obtaining bids, estimating mean WTP, constructing bid curves, aggregating data, and evaluating the CVM exercise (Hanley and Spash 1993). The CVM has been extensively used to value environmental and natural resources in developing countries (Navrud and Mungatana 1994; Mbelwa 2002). Hence, CVM was employed to value CGRs in this study. While using the CVM, the recommendations of the National Oceanic and Atmospheric Administration (NOAA) panel (Arrow et al. 1993) were followed wherever possible in designing and implementing the survey. In setting the hypothetical market for CVM, the evaluated programme was clear in its aims and consequences and the valued goods (CGRs) were well-known and of interest to the people. Although the current trend in CV surveys is towards the referendum formats using dichotomous choice (Arrow et al. 1993; Whittington 2002), the bidding game approach is also used for bid collection (Mitchell and Carson 1989; Mbelwa 2002; Dong et al. 2003). This study used an open-ended bidding game approach to elicit willingness for contribution. Two main reasons of using this approach are, first the dichotomous choice question tends to give higher estimates compared with the open ended one (Bateman et al. 1995; White and Lovett 1999), especially for studies conducted in low income countries. Second, the sample required for statistical reliability for the dichotomous choice method is higher (Mitchell and Carson 1989; White and Lovett 1999; OECD 2002). This was restricted in the present study due to time and resource constraints. The form and frequency of payment was clearly explained to the respondents. The payment vehicle in the study was an annual contribution towards the conservation programme.

The CVM is usually accompanied with doubts about its viability because of various potential biases such as hypothetical bias, information bias, strategic bias, design bias (bid vehicle, starting point), as well as embedding bias (Mitchell and Carson 1989; Hanley and Spash 1993; Hausman 1993; Mbelwa 2002). This study was designed to reduce such potential biases as far as possible. Efforts were made to deliver clear and equal information to all the respondents, the payment vehicle was clear, a pre-tested questionnaire used, and four different local enumerators were employed to reduce biases. To test the response validity, the questionnaire was designed to generate information on demographic and social background and embedded biodiversity knowledge of the respondents.

The study area

The study area is one of the in situ conservation project sites being implemented by the Nepal Agriculture Research Council (NARC), Local Initiatives for Biodiversity, Research and Development (LIBIRD), and the International Plant Genetic Resource Institute (IPGRI). It is situated in Kaski district in the mid-hill region (800 m to 1500 m) of Nepal. The topography of the region is ancient lake and river terraces in moderate to steep slopes. The area experiences high rainfall (>3900 mm per annum) and has warm temperate to subtropical climate. Mean daily minimum temperature in the coldest month is 7° and the mean daily maximum in the hottest month is 30.5° with an annual mean of 20.9°C (Rana et al. 2000). The area is reported to be a hot spot in terms of crop diversity (Rijal et al. 1998). The major crop is rice and almost all the farmers grow this crop. The farmers are maintaining 63 local landraces for this (Rijal et al. 1998; Rana et al. 1999). Nepal is the centre of origin for rice. Therefore, the study uses rice landraces as an example of CGRs.

Population, sample, and sampling

The study population was selected from 12 hamlets[1] of the project area. Proportionate random sampling was done to draw the samples by stratifying the hamlets and the socio-economic status of the people (Nichols 2000). The total samples were 20 per cent constituting 107 households drawn proportionally from three socio-economic strata.[2] As each household represented a sampling unit, one of the household members taking the decisions for agriculture was selected for the survey.

[1]Hamlets are smaller villages or are parts of the bigger village, also called *tol.*

[2]The socio-economic strata (rich, medium, and poor) were identified using the farmers' own criteria. The main parameters for the categorization were food self sufficiency of households, landholdings, government services, business, house and homestead properties, possessing house and/ or land in major cities (Rana et al. 1999; 2000). People who had more than 0.75 ha of land, were almost food self-sufficient, had permanent monthly income sources or remittances, had land and house in cities, were considered socio-economically 'rich'. Similarly, people having 0.25 to 0.75 ha of land, more than 6 months of food self-sufficiency and off-farm income sources were considered to be placed in the 'medium category' and people with less than 0.25 ha of land, less than 6 months food self-sufficiency, and working as casual labourers were categorized as 'poor'.

The survey and the questionnaire

The instrument used for the study was a survey questionnaire. Local, experienced enumerators were trained and employed to conduct the household survey. The open-ended, structured questionnaire was pre-tested with eight respondents outside the sample frame and revised before administration. The questionnaire consisted of three sections. The first section addressed the household and the respondents' demographic and socio-economic information. The second section elicited information regarding the resources (land, genetic diversity), and the perception towards general knowledge on biodiversity and particularly on agrobiodiversity and rice landraces. The third section consisted of the explanation of the study hypothesis and elicited the WTP for landraces conservation. An explanation of the subject matter was given before asking WTP questions. The respondents were briefed on the need for conservation using hypothetical market value of the landraces and were asked if they were willing to contribute for landraces conservation.

Three different methods of conservation (in situ household level, community gene bank, and ex situ or gene bank) were presented and discussed. Then, different modes of questions were used to assess the WTP for different methods of conservation. The modes of payment were: growing landraces instead of high yielding variety (HYV) on own farm, indirect payment or contribution to community gene bank, and direct cash payment for conserving their landraces in gene bank (ex situ). The question modes were as follows:

Mode 1: Are you willing to grow local landraces for the sake of conservation instead of growing your preferred variety? If yes, state the area (ropani),[3] and expected yield of those conserved varieties (n = 6) on your own land. Please provide the productivity and market price of your preferred landrace or variety.

Mode 2: Are you willing to contribute in cash or kind if you are asked for conservation of landraces in the community seed bank or gene bank? If yes, state the amount of labour (man-days), area (ropani), and inputs (chemical fertilizers, pesticides, manure, or seed).

Mode 3: Are you willing to pay cash, if you are asked to pay the government or NGOs, to conserve local landraces in an ex situ/gene bank? If yes, how much will you be willing to pay monthly? Rs 70, 60, 50, 40, 30, 25, 20, 15, 10, ...

[3]A local unit for area measurement (20 ropani = 1 hectare).

WTP bids were then calculated from the willingness to contribute using specific procedures for the respective methods of conservation.

WTP bids calculation, data analysis, and model specification

The data were entered in the Excel worksheet and WTP bids were calculated. The individuals' WTP bids for conservation of landraces in situ at the household level were calculated using the following equation. During data analysis, 14 protest bids were omitted from the data set. These bids contained negative values due to higher expected market price of the said landraces.

$$\text{WTP} = \Sigma(AiYiPi) - \sum_{i=1}^{n} (aiyipi) \qquad ...(10.1)$$

where, WTP = total willingness to pay for in situ conservation of landraces; Ai = total area for conservation = $a_1 + a_2 \cdots a_n$, Yi = yield of preferred variety; Pi= price of the preferred variety, ai = area for conservation of landraces i (i = 1to n where, n = 6); yi = expected yield of landrace i; and pi = price for the landrace i.

The individuals' WTP bids for conservation of landraces in the community gene bank were calculated using equation 10.2. None of the WTP bids was a protest bid and all the bids were used in the analysis.

$$\text{WTP} = [L * Pl + A * Pa + F + P + M * Pm + S * Ps] \; ... (10.2)$$

where, L = Labour contribution in man-days; A = area contribution in hectare; F = chemical fertilizer contribution in cash; P = chemical pesticides contribution in cash; M = manure contribution ($doko$);[4] S = seed contribution in kg; Pl = labour price; Pa = price of land renting; Pm = price of manure; and Ps = Price of seed.

In case of cash contribution for the ex situ, the WTP bids were calculated using equation 10.3. None of the WTP bids was a protest bid and all the bids were used in the analysis.

$$\text{WTP} = \left[\sum_{i=1}^{n} (Xi) / N \right] * 12 \qquad ... (10.3)$$

[4]A basket like structure, which is locally used to carry manure from the farm to the field. One $doko$ costs about Rs 20 (US$ 1 = Rs 73).

where, WTP = average willingness to pay per landrace per year; X_i = amount paid to conserve landrace i (i = 1 to n); n = number of landraces (max 6); and N = total number of landraces considered for conservation.

The WTP bids were then transferred to Minitab statistical software for analysis (MINITAB 2000). Mean WTP, SD, confidence interval (CI), and the relationship between WTP and categorical variables were analysed using descriptive statistics, 2-sample t-test, and ANOVA. The WTP bids were also regressed with various explanatory variables. The bid function was arrived at using general regression analysis, starting from all the potential explanatory variables, removing the non-significant, re-estimating the model, and so on until all remaining variables were significant at 95 per cent level (Horton et al. 2003).

The WTP function is

$$\text{WTP} = ß_0 + ß_1 X_1 + ß_2 X_2 + ß_3 X_3 + \ldots + ß_n X_n + \varepsilon_i \ldots (10.4)$$

where, WTP = farmers' willingness to contribute for landraces conservation; $ß_0$ = constant $ß_1 + ß_n$ = Coefficients; $X_1 - X_n$ = variables contributing in WTP; and ε_i = Random error $\sim N (0, 1)$.

RESULTS AND DISCUSSIONS

Sample characteristics

The demographic data obtained from the 107 households were proportionately stratified within socio-economic status as presented in Table 10.1. The mean age of the respondents was 45 years and the respondents were nearly evenly divided between males and females. Almost 46 per cent of the respondents were either illiterate or had no schooling knowledge and less than 5 per cent had attended university level education. The average household size was 6.5 which is higher than the national average at 5.44 (CBS 2003) but exactly the same as reported by Rana et al. (2000). The mean household income was US$ 1156 in the study area, which is slightly lower than the national level (CBS 2003). The average land capital of the respondents was 0.90 ha which is almost equal to the national average 0.96 ha (p>0.05) (CBS 2003; Rana et al. 2000). The average number of land parcels was 7.75 in the study area, which is very high compared to the national average of 4 (p<0.01) (CBS 2003; Rana et al. 2000). The average food self-sufficiency was 9.95 months.

Table 10.1:   Characteristics of Respondents by socio-economic strata (*n = 107*)

Characteristics	Rich	Medium	Poor	Total
Age (years, mean)	49	41	45	45
Gender (number, per cent)				
Male	22 (20.5)	23 (21.5)	8(7.5)	53 (49.5)
Female	19 (17.8)	17 (15.9)	18 (6.8)	54 (50.5)
Education of the respondents	4 (3.7)	3 (2.8)	9 (8.5)	16 (15)
(number, per cent)				
Illiterates	10 (9.3)	12 (11.3)	11 (10.4)	33 (31)
Literates	8 (7.5)	6 (5.6)	2 (1.9)	16 (15)
Primary education	10 (9.3)	14 (13.0)	4 (3.7)	28 (26)
Secondary education	7 (6.5)	3 (2.5)	0 (0)	10 (9)
Higher secondary education	2 (2)	2 (2)	0 (0)	4 (4)
University education				
Household size (mean)	7.3	5.8	6.3	6.5
Household income $ (mean)	1341	1193	820	1156
Landholdings (ha) (mean)	1.29	0.81	0.42	0.90
Parcels of Land (mean)	10.1	7.79	4.23	7.75
Food (self-sufficiency months) (mean)	12.51	9.91	6.19	9.95
Number of varieties grown (mean)	4.02	2.9	2.5	3.2
Knowledge of agrobiodiversity (number, per cent)	31 (29)	30 (28)	11 (10.3)	72 (67.3)
Yes	10 (9.3)	10 (9.3)	15 (14.1)	35 (32.7)
No				

Source: Authors.

It was found that 25 per cent of the households produced food for ≤ 6 months whereas 25 per cent produced food sufficient for ≥ 12 months. The average number of landraces (excluding modern varieties) grown was 3.2.

The findings indicate that the landholdings, annual income, literacy rate, and food sufficiency months are greater for the resource rich people as compared to those for the poor in the study site. The number of landraces grown by richer people is higher as compared to those grown by the poor.

The sample data are almost representative of the study area with respect to the gender, household size, the educational status of people, annual income, and landholdings of the people, although the average diversity of landraces is higher than in *terai* and high hills of the country (see Paudel et al. 1999; Rana et al. 1999).

Farmers' knowledge of biodiversity

Since the study was conducted in the in situ conservation project area, farmers of the study area were found to be informed about the project on the need for conservation. They were familiar with the term 'agrobiodiversity' and different methods of biodiversity conservation. Majority (67 per cent) of the farmers were familiar with the concept of agrobiodiversity. Most of the farmers explained the use value of the rice landraces relating to conservation. Although some farmers (12 per cent) perceived that landraces must be conserved for the need of future generations (intrinsic value), most of them acknowledged the need for conservation due to ecological, direct use, and option use values of landraces. This resembles the findings of Rana et al. (2000) regarding the factors influencing the conservation and maintenance of landraces on farm. Farmers were very enthusiastic in explaining the need for landraces conservation and all the respondents participated in discussing why the local rice landraces should be conserved. The farmers' perceived knowledge on the 'need for rice landraces conservation' is presented in Table 10.2.

Table 10.2:   Farmers' reasons for conservation of rice landraces in Kaski, Nepal

Need for landraces conservation	Responses (n = 107)	
	number	%
Diverse adaptability	64	59.8
Local varieties have diverse taste, use, straw need, and importance	57	53.3
Possibility of extinction from the habitat	33	30.8
Medicinal values	27	25.2
Possesses cultural value	23	21.5
These have yield stability over the years	13	12.1
Security for the future	12	11.2
Improved varieties have no guarantee of yield	10	9.3
Local landraces are suitable in adverse climatic conditions such as epidemics	8	7.5
Need for choice in seed in future	7	6.5
For improvement of new varieties	6	5.6
The basis for subsistence are landraces	2	1.8
These should be conserved because of international or global importance	1	0.9
These are being grown from ancient times	1	0.9

Source: Authors.

WTP for conservation

The people's WTP for conservation of crop genetic diversity is analysed with respect to the importance of the landraces to the farmers, the possible threat of extinction of the landraces/genetic diversity, and the different methods of conservation. Since genetic diversity, in the Nepalese context, is considered as a public good (neither having rivalry in consumption nor excludability), the analysis presents the interest of the people in the conservation of these public goods; the methods they prefer for conservation and the genetic resources that they wish to conserve.

Importance of landraces

Grain quality, market price, yield attributes, and ecological adaptability of crop landraces determine people's participation in conservation (Brush 2000), and thereby influence willingness for contribution. The direct use value of landraces mainly influences the farmers' WTP. The reasons behind the difference in willingness to contribute for crop landraces are socio-cultural importance, agro-ecological diversity, and direct consumption values (Rana et al. 2000). People's WTP has been analysed for grain quality as one of the important attributes of the rice landraces.

The landraces were categorized into three different grain qualities, based on the farmers' preferences. The basis of categorization was grain size, cooking quality, grain stickiness, and market price. The fine grain landraces are those that have very good cooking quality, high price, and are preferably aromatic in quality. The medium quality have medium cooking quality, medium price, and no aroma, while the coarse grain quality has relatively poor cooking quality and low market price. The different landraces are listed in Table 10.3.

Table 10.3:   The major rice landraces by different type of grain quality

	Types of landraces	
Fine grain quality	Medium grain quality	Coarse grain quality
Jethobudo	Ekle	Sanomadhise
Anadi	Gurdi	Madhise
Panhele	Jarneli	Manamuri
Bayarni	Gauriya	Ramsali
Jhinuwa	Flyankote	Jhauri
	Lame	Mansara

Source: Authors.

The analysis shows that the WTP for different grain quality rice landraces was different for both in situ and ex situ conservation. The mean willingness for contribution of land for in situ conservation of fine and medium grain quality landraces was significantly higher than for coarse grain (p<0.01), and for ex situ conservation, the mean WTP for fine and medium quality landraces was also higher than that for the coarse grain types (p<0.01) (Table 10.4).

Table 10.4:   The mean contribution of land (ha) for in situ conservation and mean WTP for ex situ conservation to different grain quality landraces

(WTP in US$)

	Fine grain type landraces		Medium grain type landraces		Coarse grain type landraces	
	Contribution of land for in situ (ha)	Cash ($) payment for ex situ	Contribution of land for in situ (ha)	Cash ($) payment for ex situ	Contribution of land for in situ (ha)	Cash ($) payment for ex situ
Mean	0.01275	2.337	0.01088	2.22	0.003776	1.796
95% CI	0.0096–0.016	2.1–2.6	0.0085–0.0132	2.0–2.4	0.0027–0.0049	1.5–2.1
N	107	102	107	103	107	82

Source: Authors.

The WTP for different grain quality landraces was also affected by the socio-economic status of the people (Table 10.5). Contribution of land for conservation was significantly lower for fine grain landraces by poor people compared to medium and rich people, whereas cash contribution was significantly less for medium grain landraces by the poor as compared to medium and rich, and for coarse grain both rich and poor people expressed significantly lower WTP compared to the medium income group. The reason is that poor people cannot afford to grow low yielding fine grain quality and therefore, they might have lower willingness to contribute land as compared to the rich people. Similarly, the richer people have more land to grow fine grain rice and they do not prefer the coarse grain, therefore their contribution is less. On the other hand, poor people also consider the fine grain rice landraces as important and are, therefore, willing to conserve them through a gene bank. Interestingly, poor people do not contribute more cash to conserve through gene bank because they grow landraces in their own fields.

Table 10.5:   Mean WTP for different grain quality of rice landraces by different socio-economic status of the people for in situ and ex situ conservation

	Fine grain type landraces		Medium grain type landraces		Coarse grain type landraces	
	Contribution of land for in situ (ha)	Cash ($) payment for ex situ	Contribution of land for in situ (ha)	Cash ($) payment for ex situ	Contribution of land for in situ (ha)	Cash ($) payment for ex situ
Rich	0.01810	2.471	0.01187	2.448	0.004479	1.808***
Medium	0.01103	2.388	0.01174	2.298	0.003907	2.915
Poor	0.00697**	2.035	0.00799	1.763**	0.002686	1.125***

Notes: ** p<0.05 and *** p<0.01.
Source: Authors.

Perceived threats of extinction

Analysis was conducted to know whether the perceived threat of extinction of the genetic resources influences the farmers' WTP. The landraces of rice were categorized in four cells based on the area of cultivation and number of farmers cultivating the landraces in the community (Jarvis et al. 2000; Bajracharya et al. 2003), as shown in Table 10.6. The landraces grown by many people in large or small areas, or grown by few farmers in large areas are identified as abundant landraces and these are not threatened by extinction but the landraces grown by 'few farmers in small areas' are under threat of extinction; this requires prompt action for conservation.

Table 10.6:   The landraces identified as abundant and threatened in the community[a]

Growing area	Number of Households	
	Many Farmers (>11)	Few Farmers (<11)
Large (>1.2 ha)	*Abundance* Ekle, Madhise Jethobudo, Mansara Panhele, Gurdi	*Abundance* Sano madhise
Small (< 1.2 ha)	*Abundance* Anadi, Bayarni Jarneli	*Threatened* Gauriya, Flyankote, Lame Manamuri, Ramsali, Jhauri, Jhinuwa

Source: Based on the four cell analysis suggested by Jarvis et al. (2000) and Bajracharya et al. (2003).

Farmers' WTP for conservation of rice landraces in gene bank (ex situ) showed no significant difference (p>0.10), whether there are abundant (grown by many farmers or few farmers in large area) or rare and almost at the point of extinction (grown by few farmers in small area). However, the contribution of area for conservation in situ differs significantly (p<0.01). The area contribution was higher for those landraces, which are at present grown by many farmers in larger area than for the other three categories (Table 10.7).

Table 10.7:   The mean contribution of land (ha) for in situ conservation and mean WTP for ex situ conservation for abundant and threatened rice landraces in the community[a]

	Many Farmers (>11)		Few Farmers (<11)	
	Area (ha) for in situ	Cash ($) for ex situ	Area (ha) for in situ	Cash ($) for ex situ
Large Area (>1.2 ha)				
Mean	0.01783	2.30	0.007	2.00
95% CI	0.0138–0.0219	2.07–2.52	0.005–0.009	1.77–2.24
N	94	98	85	85
Small Area (< 1.2 ha)				
Mean	0.00407	2.13	0.00643	2.01
95% CI	0.0033–0.0048	1.89–2.38	0.0037–0.009	1.65–2.37
N	94	99	42	40

(WTP in US$)

Source: Authors.

The analysis shows that farmers are not willing to contribute to conserve CGRs which are rare and prone to extinction. Their strategy for contribution is more influenced by the importance of use values (cultural, social, and ecological) than the non-use or option (intrinsic or existence) values.

Methods of conservation

The different methods of conservation influence the people's WTP for CGRs. Three different methods of conservation (community gene bank, in situ at the household level, and ex situ) were discussed with farmers and they were asked for their contributions towards conservation.

Mean WTP per annum: The mean willingness to contribute to conserve genetic resources, particularly the rice landraces by three different methods of conservation was significantly different (p<0.05). The highest WTP was for conservation through a community bank, where all the farmers enthusiastically agreed to participate and provide either land, labour, or cash, or all the inputs. The mean WTP for community conservation was four times higher than that for in situ household level conservation and eight times higher than for ex situ conservation (Table 10.8). This means that farmers perceived the community gene bank conservation as the appropriate method for CGR conservation. Through in situ community conservation, farmers not only conserve but also utilize and have access and control over all the CGRs.

Table 10.8:   Mean WTP per Annum per household for landrace conservation for three different methods

	In situ at household level	Community gene bank	Ex situ or gene bank
Mean	US$ 4.181	US$ 16.75	US$ 2.195
95% CI	US$ 2.832–5.530	US$ 13.74–19.75	US$ 1.994–2.396
N	93	107	107

Source: Authors.

Factors influencing the contribution: Table 10.9 shows that household size is significantly associated with the WTP bids for ex situ conservation. The larger households are willing to contribute less, which is consistent with economic theory because with a higher number of members more cash is needed for running the household, which decreases the capacity to contribute towards conservation or similar purposes. So, they might not be willing to contribute for ex situ conservation of landraces, instead they can contribute more to in situ conservation. Age of respondent was only associated with the in situ community gene bank conservation where elderly people are willing to contribute more than the younger population. Household income and food-sufficiency status of the family was not significantly associated with the WTP bids. Although this differs from what one would expect based on economic theory, the empirical result by the World Bank (1993) on WTP for quality water supply also does not depend on income. Similarly, it has been found by Aldy et al. (1999)

that the low income population has disproportionately higher environmental risk than the wealthy or high income population. The conservation of landraces is, therefore, more important to those who depend on agriculture and have low income. This might be the reason that income and food-sufficiency has no effect on the WTP for landraces conservation. Larger landholdings are associated with higher WTP bids, which is theoretically acceptable because people having more land are richer people who have more need of such landraces in the future when they expect to change the varieties or landraces in their diverse lands (Rana et al. 2000). Although more land parcels require a higher number of landraces (Rana et al. 2000), the WTP was not associated with the number of land parcels. Number of landraces grown was negatively associated with the WTP bids. An increase in the number of landraces grown, decreased the contribution for conservation which is theoretically acceptable because when farmers already have a large number of landraces on

Table 10.9: Full and best regression models of explanatory variables determining WTP for in situ at household level, community gene bank, and ex situ conservation of rice landraces

Explanatory	All variables included in the model			Best Fitting model		
variables	In situ at Household level	Community bank	Ex situ	In situ at Household level	Community bank	Ex situ
Constant	0.321	−5.26	2.47**	−0.63	2.20	2.254***
Household size	0.403	−0.924	−0.069*	–	−1.17**	−0.093***
Respondent age	−0.079	0.324**	−0.005	–	0.256**	–
Education level	0.867*	1.337	−0.109	1.01**	–	–
Household income (US$)	−1.2x10⁻⁴	6.4x10⁻⁴	9x10⁻⁵	–	–	–
Food-sufficiency	0.620**	0.395	0.043	0.63***	–	–
Landholding	0.569	10.4***	0.566*	–	11.73***	0.616***
Land parcels	−0.046	−0.416	0.019	–	–	–
Number of landraces grown	−1.04**	1.337	−0.116*	−1.06***	–	–
R^2	0.28	0.29	0.22	0.25	0.26	0.1
N	93	107	107	93	107	107

Notes: * $p < 0.10$, ** $p < 0.05$, and *** $p < 0.01$.
Source: Authors.

their farm, they are not interested in paying for conservation of the same landraces.

The mean WTP was higher for male respondents, the richer households, and the people having previous knowledge of biodiversity (Table 10.10). The mean WTP was significantly higher for male respondents for in situ conservation methods, but not for ex situ conservation. The mean WTP for the resource rich people was significantly higher than that for the resource poor people for all methods of conservation.

Table 10.10: The categorical variables influencing the farmers' willingness to contribute (US$) for rice landraces conservation in Kaski, Nepal

Influencing variables	Mean willingness to contribute for rice landraces conservation		
	In situ at household level	Community bank	Ex situ
Sex of the respondents			
Male	5.711**	20.48**	2.328
Female	2.618	13.08	2.064
Socio-economic status			
Resource rich	6.158***	20.44***	2.317*
Resource medium	6.079	19.20	2.330
Resource poor	0.827	7.15	1.794
Knowledge of biodiversity			
Rich knowledge	5.331**	18.18	2.289
Poor knowledge	1.644	13.81	2.002

Notes: * p<0.10, ** p<0.05, and ***p<0.01.
Source: Authors.

CONCLUDING REMARKS

The mean willingness to contribute for conservation, either in situ or ex situ for important CGRs such as rice landraces was higher than for the less important ones. The use values of the concerned genetic resources were considered as more important than non-use values because the grain quality at present is more important than the resistance gene for future. Similarly, the willingness to contribute to conservation of CGRs that are threatened by extinction was lower than that for the abundant landraces. Landraces that are considered by farmers to be less important should, therefore, be conserved ex situ or in gene banks but the CGRs that farmers consider important should be conserved in situ.

The mean WTP for community conservation was higher than in situ at household level and ex situ conservation. The difference was mainly because such goods are also public goods (Bräuer 2003) having neither rivalry in consumption nor excludability. However, this significant difference of WTP for different methods of conservation does not mean that farmers have valued the resources differently but that they have elicited the best method for conservation. This WTP denotes that at least some amount of the conservation costs could be covered by the farmers who are the users of the genetic resource. How to fund the conservation of genetic resources is, however, a political rather than a scientific question.

The better conservation approach for landraces that are highly valued by farmers seems to be the in situ conservation and best approach is the in situ community conservation. We do not argue that the CVM is a panacea in solving the problem of decision making concerning genetic resource conservation but simply put forward the point that CV should be considered as a crude guide while implementing conservation strategies. Such studies conducted for valuation of agrobiodiversity will open the door for further research and valuation of agricultural resources including horticultural and floricultural crops.

REFERENCES

Aldy, J.E., R.A. Kramer, and T.P. Holmes (1999), 'Environmental Equity and Conservation of Unique Ecosystems: An Analysis of the Distribution of the Benefits for Protecting Southern Applachian Sprucefir Forests', *Society and Natural Resources*, Vol. 12, pp. 93–106.

Arrow, K.J., R. Solow, E. Leamer, P. Portney, R. Radner, and H. Schuman (1993), *Report of the NOAA Panel on Contingent Valuation.* 58 Fed. Reg. 4601, Government Printing Office, Washington DC.

Bajracharya, J., P. Tiwari, A. Subedi, R.K. Tiwari, Y. Pandey, D. Pandey, R. Yadav, Rijal, D., B. Baniya, M. Upadhyay, B. Sthapit, and D. Jarvis (2003), 'Assessment of Local Crop Genetic Diversity in Nepal', in D. Gauchan, B. Sthapit, and D. Jarvis (eds), *Agrobiodiversity conservation on farm: Nepal's contribution to scientific basis for national policy recommendations.* IPGRI, Rome, Italy.

Bateman, I.J., I.H. Langford, R.K. Turner, K.G. Willis, and G.D. Garrod (1995), 'Elicitation and Truncation Effects in Contingent Valuation Studies', *Ecological Economics*, Vol. 12, pp. 161–79.

Bräuer, I. (2003), 'Money As an Indicator: To Make Use of Economic Evaluation for Biodiversity Conservation', *Agriculture, Ecosystems and Environment*, Vol. 98, pp. 483–91.

Brush, S.B. (2000), 'The Issue of *in situ* Conservation of Crop Genetic Resources', in S.B. Brush (ed.), *Genes in the Field: On Farm Conservation of Crop Diversity.* IPGRI, Rome, Italy/IDRC, Ottawa Canada/Lewis Publisher.

CBS (2003), *Statistical Year Book of Nepal*, Central Bureau of Statistics, Kathmandu, Nepal.

Dong, H., B. Kouyate, J. Carins, and R. Sauerborn (2003), 'A Comparison of the Reliability of the Take-it-or-leave-it and the Bidding Game Approaches to Estimating Willingness to Pay in a Rural Population in West Africa', *Social Science and Medicine*, Vol. 56, pp. 2181–9.

Gauchan, D. (1999), 'Economic Valuation of Rice Landraces Diversity: A Case Study of Bara ecosite, Terai, Nepal', in B. Sthapit, M. Upadhyay, and A. Subedi (eds.), *A Scientific Basis of in situ Conservation of Agrobiodiversity on Farm: Nepal's contribution to global project*, NP Working paper No. 1/99. NARC/LIBIRD, Nepal/IPGRI, Rome Italy.

Gupta, S.R., M.P. Upadhyaya, and T. Katsumo (1996), 'Status of Plant Genetic Resources of Nepal', 19th summer crop workshop. RARS Parwanipur, Nepal.

Hammer, K., T. Gladis, and A. Diederichsen (2003), '*In situ* and on-farm Management of Plant Genetic Resources, *European Journal of Agronomy*, Vol. 19, No. 4, pp. 509–17.

Hanley, N. and C.L. Spash (1993), *Cost Benefit Analysis and the Environment*, Edward Elgar Publishing Limited, United Kingdom.

Hausman, J. (1993), *Contingent Valuation—A Critical Assessment*: North-Holland, Amsterdam.

Horton, B., G. Colarullo, I.J. Bateman, and C.A. Peres (2003), 'Evaluating Non-user Willingness to Pay for a Large-scale Conservation Programme in Amazonia: A UK/Italian contingent valuation study', *Environmental Conservation*, Vol. 30, No. 2, pp. 139–46.

Jarvis, D., B. Sthapit, and L. Sears (2000), 'Conserving Agriculture Biodiversity *in situ:* A Scientific Basis for Sustainable Agriculture', International Plant Genetic Resources Institute, Rome, Italy.

Mbelwa, J.R. (2002), 'Improving Beach Management: An Analysis of the Role of the Government and Local Community in the Management of

8 PAYMENT FOR ECOSYSTEM SERVICES

the Beach Areas in Dar es Salaam', *NORAGRIC,* MSc Thesis. Agricultural
University of Norway.
MINITAB (2000), *MINITAB for Windows: MINITAB Statistical Software,*
MINITAB Inc.
Mitchell, R.C. and R.T. Carson (1989), *Using Surveys to Value Public
Goods: The Contingent Valuation Method,* Resource for the Future,
Washington DC.
Navrud, S. and E.D. Mungatana (1994), 'Environmental Valuation in
Developing Countries: The Recreational Value of Wildlife Viewing',
Ecological Economics, Vol. 11, pp. 135–51.
Nichols, P. (2000), *Social Survey Methods: A Field Guide for Development
Workers,* Oxfam Publishing, Oxford, UK.
Nunes, P. and J.C.J.M. Van den Bergh (2001), 'Economic Valuation of
Biodiversity: Sense or nonsense?' *Ecological Economics,* Vol. 39, No. 2,
pp. 203–22.
OECD (2002), *Handbook of Biodiversity Valuation: A Guide for Policymakers,*
OECD Publications, Paris, France.
Paudel, C.L., P. Chaudhary, D.K. Rijal, S.N. Vaidya, P.R. Tiwari, R.B. Rana,
D. Gauchan, S.P. Khatiwada, S.R. Gupta, and B.R. Sthapit (1999), 'Agro-
ecosystem Factors for *in situ* Conservation of Agrobiodiversity in Different
Eco-zones of Nepal', in B. Sthapit, M. Upadhayay and A. Subedi (eds.),
*A Scientific Basis of in situ Conservation of Agrobiodiversity on Farm: Nepal's
contribution to global project.* NP Working paper No. 1/99. NARC/
LIBIRD, Nepal /IPGRI, Rome, Italy.
Rana, R.B., P. Chaudhary, R.K. Tiwari, D.K. Rijal, A. Subedi, D. Jarvis,
and B.R. Sthapit (2003), 'Strengthening Community Based Organizations
for Effective On-farm Conservation: Some Lessons and Issues from the
in situ Project Nepal', in B.R. Sthapit, M.P. Upadhayay, B.K. Baniya,
A. Subedi, and B.K. Joshi (eds), On-farm Management of Agricultural
Biodiversity in Nepal, Proceedings of a National Workshop, Lumle Nepal.
NARC, LI-BIRD and IPGRI.
Rana, R.B., D. Gauchan, D.K. Rijal, S.P. Khatiwada, C.L. Paudel,
P. Chaudhary, and P.R. Tiwari (1999), 'Cultural and Socio-economic
Factors Influencing Farmer Management of Local Crop Diversity:
Experience from Nepal', in B. Sthapit, M. Upadhayay, and A. Subedi
(eds.), *A Scientific Basis of in situ Conservation of Agrobiodiversity on
Farm: Nepal's contribution to global project,* NP Working paper No.
1/99. NARC/LIBIRD, Nepal/IPGRI, Rome Italy.
Rana, R.B., D.K. Rijal, D. Gauchan, B. Sthapit, A. Subedi, M.P. Upadhyay,
Y.R. Pandey, and D.I. Jarvis, (2000), *In situ Crop Conservation: Findings
of Agro-ecological, Crop Diversity and Socio-economic Baseline Survey of
Begnas Ecosite Kaski, Nepal,* NP Working paper No. 2/2000. NARC/
LIBIRD, Nepal /IPGRI, Rome Italy.

Rijal, D.K. (2002), 'Concerning the Biodiversity Conservation, the Concept of community seed banking: Process and Constraints', Adarsha Samaj. Pokhara.
Rijal, D.K., R.B. Rana, K.K. Sherchand, B.R. Sthapit, Y.R. Pandey, N. Adhikari, K.B. Kadayat, Y.P. Gauchan, P. Chaudhary, C.L. Poudel, S.R. Gupta, and P.R. Tiwari (1998), *Strengthening the Scientific Basis of in situ Conservation of Agrobiodiversity: Finding of Site Selection Kaski, Nepal,* NP Working paper No. 2/1998. NARC/LIBIRD, Nepal /IPGRI, Rome Italy.
Sthapit, B. and D. Jarvis (2003), 'Implementation of On-farm Conservation in Nepal', in D. Gauchan, B. Sthapit, and D. Jarvis (eds), *Agrobiodiversity Conservation on Farm: Nepal's Contribution to Scientific Basis for National Policy Recommendation,* IPGRI, Rome, Italy.
Sthapit, B.R., M.P. Upadhayay, B.K. Baniya, A. Subedi, and B.K. Joshi (2003), *On-farm Management of Agricultural Biodiversity in Nepal,* Proceedings of a National Workshop, Lumle Nepal, NARC, LI-BIRD and IPGRI.
Turpie, J.K. (2003), 'The Existence Value of Biodiversity in South Africa: How Interest, Experience, Knowledge, Income, and Perceived Level of Threat Influence Local Willingness to Pay', *Ecological Economics,* Vol. 46, No. 2, pp. 199–216.
UNEP (1994), *Convention on Biological Diversity: Text and Annexes UNEP/ CBD/94/1, The Interim Secretariat for the Convention on Biological Diversity,* Geneva.
UNEP (1995), *Global Biodiversity Assessment,* Cambridge University Press.
White, P.C.L. and J.C. Lovett (1999), 'Public Preferences and Willingness to Pay for Nature Conservation in the North York Moors National Park, UK', *Journal of Environmental Management,* Vol. 55, pp. 1–13.
White, P.C.L., A.C. Bennett, and E.J.V. Hayes (2001), 'The Use of Willingness to Pay Approaches in Mammal Conservation', *Mammal Review,* Vol. 31, No. 2, pp. 151–67.
Whittington, D. (2002), 'Improving the Performance of Contingent Valuation Studies in Developing Countries, *Environmental and Resource Economics,* Vol. 22, pp. 323–67.
Worede, M. (1992), Ethiopia: 'A Gene Bank Working with Farmers', in D. Cooper, R. Vellve, and H. Hobbelink (eds), *Growing Diversity: Genetic Resources and Local Food Security,* Intermediate Technology Publications.
World Bank (1993), *World Development Report,* Oxford University Press, New York.

11

A Travel Cost Analysis of Non-market Benefits of Forest Recreation in Great Britain

AYELE GELAN

INTRODUCTION

In recent decades, there has been a considerable increase in the number of people taking part in forest recreation in most developed economies. In the UK, the Forestry Commission periodically reports visitor numbers on its own woodland (NAO 1986; Benson and Willis 1992). Their estimates ranged from 24 to 32 million in 1980s. In the 1990s, this figure increased to about 50 million (Christie et al. 2005). This means that the total number of visitors to forestry commission woodlands nearly doubled during the decade. Similarly, a regular survey of day visits to forests in the UK, known as UKDVS, indicated that visitor numbers to all woodlands in the UK increased by 4.3 per cent per annum between 1994 and 1998 (Christie et al 2005, p. 10). This suggested that a significant shift has been taking place in public taste in favour of forest recreation and related goods and services.

In order to meet the rapidly rising public demand for access to forest recreation, the Forestry Commission has been striving to increase opportunities for public recreation within its forests and also promote these benefits in private woodlands. Improving facilities for public access to forest recreation has involved heavy public investment but such use of tax payer money would require justification and evidence-based decision making. To that effect, the Commission has sponsored numerous valuation studies to assess the impact of the different

The author would like to acknowledge that this paper was written and submitted while he was a research fellow at the Macaulay Land-use Research Institute, Aberdeen, UK.

forms of forest recreation in Great Britain (Christie et al. 2006; Christie et al. 2005; Hill et al. 2003; Scarpa 2003; Willis et al. 2003; Willis et al. 2000; Willis et al. 1989).

There are differences in the sizes of estimates of forest recreation benefits provided by different studies. Most studies set out with the aim of estimating economic surplus generated in the process of forest recreation. This represents the sum of non-market benefits associated with forest recreation. Christie et al. (2005) and Leeworthy and Bowker (1997) discuss the importance of distinguishing between different concepts in undertaking valuations of non-market benefits of natural resources such as benefits from forest recreation. These may include both 'use' and 'passive use' value elements. While use values relate to the utility gained by an individual from direct participation in a forest recreation activity, passive-use values could be less obvious; they may relate to option values, which relate to the utility gained by an individual for the option to use forest recreation in the future, or vicarious values, the utility gained from the knowledge that the recreation resource is maintained for others to use; or bequest values, which could be left for future generations to use. The non-market benefits from forest recreation may involve some or all of these values.

Since market prices do not exist for most forest recreation goods and services, alternative economic valuation techniques have been developed to measure these values. Zawacki et al. (2000) classified the non-market valuation methods into two basic categories, those that rank outcomes and those that indirectly estimate monetary values. Conjoint analysis is one ranking and survey-based method that measures the joint effect of two or more product or service attributes on consumer preferences in utility units, for example, accommodation, fees, boat launch access, and wildlife habitat (Zinkhan et al. 1997 and Dennis 1998). The two most popular valuation methods used to estimate monetary values for non-market goods are the CVM and the TCM. What these two methods have in common is the fact that they estimate CS or net WTP— the amount by which benefits from a visit to a forest would exceed the travel cost incurred by the visitor. In the absence of a market price, CS has become increasingly acceptable for use in benefit–cost calculations required for economic efficiency analyses.

In this study, the TCM is used with data from on-site survey visitors at 44 different forest sites throughout the UK. Since the TCM is based on actual behaviour or a RP of a visitor, it is believed that it would enable one to overcome inherent weaknesses of SP-based methods such as CVM (Scarpa 2003). For instance, the validity of CVM results has been questioned (Zawacki et al. 2000) because of its hypothetical nature in extracting information from individuals. On the contrary, the TC technique relies on establishing a relationship between the costs incurred by a visitor to a forest site and the frequency of such visits taken per annum. This relationship is used to derive CS for accessing a forest recreation site. TCM has been used extensively in forest-related recreation research to value site access as well as changes in site quality (Englin et al. 2003; Willis and Garrod et al. 1992; Englin et al. 1995). However, TC has its own shortcomings; the main one being the fact that it is limited to on-site sample surveys.

The main objective of this study is to estimate per person-trip and aggregate CS from accessing forest recreation in the UK. Although there have been numerous application of TCM to UK data, considerably large variations exist, particularly in the sizes of per person-trip CS reported. In general, there is a lack of methodological clarity and hence it is not possible to judge as to why and how such substantial differences in estimation results could arise. As noted earlier, there is a good deal of demand from policy makers for credible and reliable estimates of public benefit from forest recreation. This study is believed to make a modest contribution to methodological clarity in TCM applications in the UK.

The remaining part of the chapter is structured as follows. The next section discusses data sources and transformation methods. The third section highlights the empirical model, presenting its equations and then defining selected model variables. The fourth section presents econometric results, calculates CS, and then compares the outcome of this study with those obtained from previous studies in forest recreation in the UK. The final section provides concluding remarks.

DATA SOURCES

The primary source of data for this study comes from a survey of forest visitors in the UK. The survey was undertaken to implement a study commissioned by the Forestry Commission to provide greater understanding of the role of forests in tourism. The study was conducted by a team of researchers, including the author of this research work, from the Macaulay Land Use Research Institute and

University of Gloucestershire. The survey was carried out during July, August, and September 2002 at 44 forest sites located throughout England, Scotland, and Wales. 30 of the 44 woodland sites were owned by the National Forest Enterprise; 7 sites by the Woodland Trust and 8 by the Royal Society for Protection of Birds (RSPB).

Interviews were conducted at entrance or exit points of each forest or woodland site. Each site was assigned one interviewer who selected respondents for interview on a continuous basis: approaching the next person to pass the interviewer after completing the previous interview. A total of 1906 such face-to-face interviews were conducted with adults over 16 years old. For visitors in a group, one person was selected for interview. A quota of 45 interviews was set at each site, although this quota was not achieved at a few sites where visitors were particularly sparse.

The questionnaire was structured to collect quantitative and qualitative information on various aspects of forest recreation including visit characteristics (for example, day trips or overnight stays, motivation for the trip), expenditure on different categories (such as travel cost and food), and socio-economic characteristics of visitors (age, sex, income), and attitudes of visitors towards nature conservation. Hill et al. (2003) provide further details of the study and analysis of data collected. The current study focuses on key variables that are relevant for TCMs to determine the size of CS generated due to forest recreation. Detailed discussion of variables selected for the purpose of this study is provided in the next section.

In addition to data collected on-site from forest visitors, the implementation of TCM in the context of multi-site locations requires forest specific data such as existence of water features in the forest area; size of population within certain perimeters of the forest sites; the area of the forest site; and trail lengths in the forest for forest walks. This study benefited from the existence of such an additional database which was also created in the process of implementing the Forestry Commission study (Hill et al. 2003).

EMPIRICAL MODEL

Model equations

The specification of a TC demand function is most commonly modelled with a single equation to obtain aggregate demand as the sum of individual values calculated from a count data model (Zawacki et al. 2000). The values to be aggregated are represented by the CS that is obtained as the integral of the area under the individual's demand curve and above the price line. The logic of summing over individual demand to obtain total demand for recreation at a single site can be extended to estimating aggregate national demand from multiple sites (Englin et al. 2003).

In this study, a single equation is employed to estimate the national demand for recreation trips, and the general specification of demand for recreation trips is:

$$Q_{ij} = f(TC_{ij}, SM_{ij}, SE_{ij}) \qquad \ldots (11.1)$$

where Q_{ij} is the number of trips by the ith individual to site j; TC_{ij} is out-of-pocket money cost of the individual's ith trip to site j; OT_{ij} is a vector of other travel related variables including travel time; SM_{ij} is a dummy variable denoting whether individual's ith trip to jth site is single or multi-purpose; and SE_{ij} is a vector of socio-economic variables for individual i. Early experiments with the data indicated the presence of over-dispersion in the data and hence the Poisson Model is not suitable to model this data. A Negative Binomial (NEGBIN) Model, which corrects for data over-dispersion, was an obvious choice for this study. Following Yen and Adamowicz (1993), a standard negative binomial probability distribution can be represented as:

$$P(Q_i = q_i; q_i = 0, 1, 2 \cdots n) = \frac{\Gamma\left(q_i + \dfrac{1}{\alpha_i}\right)}{\Gamma(q_i + 1)\Gamma\left(\dfrac{1}{\alpha_i}\right)}$$

$$(\alpha_i - \mu_i)^{q_i}(1 + \alpha_i \mu_i)^{-\left(q_i + \frac{1}{\alpha_i}\right)} \qquad \ldots (11.2)$$

where $\mu_i = \exp(\beta, TC_i, OT_i, SM_i, SE_i)$; b is a vector of coefficients; Γ represents the gamma function; α is the over-dispersion parameter.

Two versions of the negative binomial model are considered for this study. The first model accounts for the fact that the data is zero-truncated, that is there are no zero counts since the survey was undertaken on-site. In other words, only those who happened to be on the site have been sampled but this is incompatible with the functional form for a standard NEGBIN distribution, which incorporates zero counts (see equation 11.2). Consequently, a correction needs to be made for this sampling bias by dropping zero counts. Following Zawacki et al (2000), a functional form for a NEGBIN model (hereafter MODEL 1), which accounts for zero-truncated nature of on-site survey data, is implemented. This is given as:

$$P(Q_i = q_i; q_i = 1, 2 \cdots n)$$

$$= \frac{\Gamma\left(q_i + \dfrac{1}{\alpha_i}\right)(\alpha_i \mu_i)^{q_i}(1 + \alpha_i \mu_i)^{-\left(q_i + \frac{1}{\alpha_i}\right)}}{\Gamma(q_i + 1)\ \Gamma\left(\dfrac{1}{\alpha_i}\right)\Pr(q_i > 0)} \qquad \dots (11.3)$$

where $P(q_i = 0) = (1 + \alpha_i \mu_i)^{-\left(\frac{1}{\alpha_i}\right)}$;

$$P(q_i > 0) = 1 - (1 + \alpha_i \mu_i)^{-\left(\frac{1}{\alpha_i}\right)}.$$

The second version of the NEGBIN model we have implemented in this study accounts for the effect of endogenous stratification, which essentially means on-site sampling does not only have a bias because only visitors were interviewed but also a good proportion of those visitors are likely to be enthusiastic users of forest recreation facilities. In such circumstances, Englin et al. (2003) provide a modified functional form (hereafter MODEL 2), which simultaneously takes account of zero-truncation as well as endogenous stratification of on-site sample-based count data:

$$P(Q_i = q_i; q_i = 1, 2 \cdots n)$$

$$= \frac{\Gamma\left(q_i + \dfrac{1}{\alpha_i}\right)}{\Gamma(q_i + 1)\ \Gamma\left(\dfrac{1}{\alpha_i}\right)} \alpha_i^q \mu_i^{(q_i - 1)}(1 + \alpha_i \mu_i)^{-\left(q_i + \frac{1}{\alpha_i}\right)} \qquad \dots (11.4)$$

Thus, two sets of econometric results were estimated in this study, one using Model 1 and the other using Model 2.

Variable Definitions

The empirical model was implemented by including variables which are defined in Table 11.1. The dependent variable, SITEVPPY, is the average number of trips to the forest site per person per year. This was calculated by multiplying the average number of trips reported by the respondent and party size, the number of people in the group. This type of variable transformation has been used in many studies because group travel by car is common practice for different types of day trip recreations including forest visits (Bhat 2003; Bowker et al. 1996; Leeworthy and Bowker 1997; Ward and Loomis 1986).

Five independent variables are most directly related to the decision to travel to a forest site. As in all TC models, the most important explanatory variable is the cost of travel to a forest site, TRAVELCOST. It denotes out-of-pocket monetary cost per person per recreation trip obtained by dividing total monetary cost reported by the number of people in the group or party-size. In some models, the opportunity cost of travel time is converted to monetary equivalent and then added to out-of-pocket money cost. However, this has remained the most controversial element in the TC literature (Martinez-Espineira and Amoako-Tuffour 2005; Englin et al. 2003). Leeworthy and Bowker (1997) discuss problems associated with converting opportunity cost of time to monetary equivalents in TC models. To begin with, there is no consensus whether to account for only round trip travel time (TRAVELTIME) or also include time spent on-site during forest recreation. The next hurdle is the actual calculation of opportunity cost of time. The most common approach has been to use the hourly wage rate of the individual and then apply this to the number of travel hours reported or estimated.

However, this implies perfect substitutability between work and leisure time, which is unrealistic because people may not find it easy to convert leisure time to earnings at normal or overtime rate by simply opting for work rather than leisure. In fact, Leeworthy and Bowker (1997) found that about 85 per cent of those surveyed reported that they lost no opportunity to earn income during their recreational visits. Thus, I avoid the arbitrary practice of converting travel time and on-site time into earnings forgone at full or percentage of the prevailing wage rate. Instead, the reported round trip time (TRAVELTIME) and the amount time spent on-site (SITEHRS) entered the empirical model as separate variables to indicate behavioural response to travel duration itself, without necessarily linking the latter to income loss. Two additional dummy variables related to travel behaviour entered the model: TRIPTYPE, to distinguish between day trips and overnight trips; and SITEONLY

Table 11.1: Definition of model variables

Variables	Definition
SITEVPPY	Annual trips from home to forest site (dependent variable)
TRAVELCOST	Visitor's out of pocket round trip travel cost to the forest site (£)
TRAVELTIME	Round trip travel time (in hours) to the forest site (log of)
SITEHRS	Number of hours spent on forest site
TRIPTYPE	Dummy: day trip = 1, otherwise = 0
SITEONLY	Dummy: 1 if visiting site only, 0 otherwise
INCOME	Monthly income (log)
AGE	Respondents age (log)
SEX	Dummy: male = 1, female = 0
HHOLDSIZE	Number of individuals in the respondent's household size
EDU	Dummy: 1 if university education, 0 otherwise
FREEDAYS	Respondent's discretionary free days per annum (log)
POP2HR	Population size within 2 hours drive from the forest site (log)
FORSIZE	Forest area in ha (log)
TRAILEN	Maximum length of walking trails (in km) the forest site (log)
WATER	Dummy: 1 if water features at forest site, 0 otherwise

Source: Authors.

denoting whether the trip is made only to visit the site or in combination with other activities such as part of a holiday in the region or further trips to other forest sites.

Four socio-economic variables were included in the TC model. In order to simplify the computational process, the natural logarithm of household income is used to obtain the variable INCOME. The influence of income in TC models has often been weak. While some studies found that income has a negative or non-significant effect (Loomis 2003; Sohngen et al. 2000; Creel and Loomis 1990), other studies found a significant positive effect (Bin et al. 2005). Visits to a forest site could also be considered as an inferior good (Liston-Heyes and Heyes 1999).

Similarly, AGE is given as the logarithm of respondent's age. Number of individuals in the respondent's household and gender of the respondent are denoted by variables HHOLDSIZE and SEX, respectively. The educational level of the respondent entered the model as a dummy variable with unity for university education and zero otherwise. It is expected that the level of educational attainment would have a positive effect on recreation trips but Shrestha (et al. 2002) found a negative effect of the same on fishing trips.

It is important to account for the effect of discretionary free time the respondent has per annum. This is represented by the variable FREEDAYS, which is a logarithm of calculated amount of free days per year. The latter is computed from the employment status of surveyed visitors (for example, employed—full time or part time, unemployed, or retired). It is assumed that there are 220 working days per year in Great Britain, which means a person employed full-time is assumed to have 145 free days (365 less 220). A retired or unemployed person, on the other hand, is assumed to have 365 free days. It is important to note that larger free time should not necessarily translate into more forest visits per annum in that those who have more free time, for example, the unemployed, are less likely to be able to afford to pay for travel to visit sites further away.

In addition to survey data, the TC model was implemented with data collected on attributes of individual forest sites. POP2HR represents population size in the immediate surroundings of the sites within two hours drive time. FORSIZE stands for total land area of the forest site in hectares. TRAILEN is the trail length created for forest walk. Finally, WATER captures whether or not there is a water feature in the forest site.

Although the survey covered a total of 1906 respondents, only 507 visitors responded consistently to all model variables. This means that the final sample utilized for the estimation of model results was 507, a great proportion of the questionnaire being discarded due to non-response to one or more of the model variables.

Econometric estimation results

The empirical estimation was undertaken using a negative binomial model (NEGBIN). Table 11.2 provides a summary of econometric results, which best fitted the data. Results under MODEL 1 account for truncated nature of the data but not its endogenous stratification while those under MODEL 2 account for both truncation and endogenous stratification. The estimation was undertaken using SAS software. Given that SAS does not provide a standard computational procedure to fit truncated and endogenously stratified data for NEGBIN models, these features were programmed and solved using a non-linear mixed model (Englin et al. 2003). The latter is a suitable tool to fit truncated and endogenously stratified count data to a NEGBIN model.

Table 11.2:   Non-linear mixed forest recreation demand model results

	Model 1	Model 2
INTERCEPT	–0.8666	–1.1810
Travel-related variables:		
TRAVELCOST	-0.1196 *	–0.1449 *
TRAVELTIME	-0.4536 *	–0.4884 *
TRIPTYPE	0.9232 *	1.0274
SITEHR	–0.1098 *	–0.1211
SITEONLY	0.1682	0.2087
Socio-economic variables:		
INCOME	0.2908 *	0.3172 *
AGE	0.3335 **	0.3485 **
SEX	0.2824 *	0.2789 *
HHOLDSIZE	0.01124	0.01419
EDU	–0.05934	–0.05492
FREEDAYS	-0.2613 **	–0.3021 **
Site-specific variables:		
FORSIZE	0.04541	0.06048 **
POP2HR	0.01561	0.01254
WATER	0.07041	0.1202
TRAILEN	–0.01034	–0.01267
N	507	507
Consumer surplus	8.36	6.90
Log-likelihood	–3017	–2018
pseudo-R²	0.78	0.81
Akaike Information Criteria (AIC)	6068.4	4069.9

Notes: * $|t| < 0.01$; **: $|t| < 0.05$
Source: Author.

It is useful to begin by focusing on the lower part of Table 11.2 where two sets of information related to the model results are provided. The first two are additional information. The calculation of CS is separately discussed but at the moment it suffices to say that Model 2 gives a slightly smaller and perhaps more realistic estimate than Model 1. The second set of information in the lower part of Table 11.2 refers to model fit statistics. The log-likelihoods are computed as –3017 and –2018 for models 1 and 2 respectively. These are used to compute the pseudo-R² statistics for each model, whereby the corresponding pseudo-R² is equal to 1 less (L1/L0), where L0 and L1 are the intercept-only and full model log-likelihoods, respectively.

The pseudo-R^2 for model 1 is 0.78 and that of model 2 is 0.81. This indicates that while both models fit the data reasonably well, Model 2 performs better than Model 1. The AIC is computed as $-2\ln(L) + 2 k$, where L is the likelihood function and k is the number of free parameters, which is 17 in our estimation (the intercept and the other covariates listed in first column of Table 11.2). The AIC statistic is mostly used to compare competing models fit. The model with the smaller AIC is said to fit the data better. Accordingly, model 2 fits the data better than model 1. Similarly, in each case, the AIC statistic for the full model is considerably smaller than the corresponding figure for the intercept-only or the restricted model (reduced by 80 per cent), which means that the full model has explained the effects of the covariates on the dependent variable.

Having highlighted the key model fit statistics, we now turn our attention to details of the effects of each explanatory variable. For all variables, the econometric results show consistently similar signs under both models. However, in almost all cases, the absolute values of coefficients under model 2 are consistently higher than those under model 1. Zawacki et al. (2000, p. 17) obtained similar results and stated that 'the size of the coefficient [of travel cost] is remarkably similar across all negative binomial models, although it rises in absolute value, as expected, when the specification adjusts for over-sampling of avid users'.

TRAVELCOST and TRAVELTIME have the expected signs and negative coefficients, indicating an inverse relationship between number of trips to forest sites and out-of-pocket travel cost as well as the number of hours taken to travel to a forest site, respectively. In other words, the further away a visitor lives from a forest recreation site the fewer number of trips per year she/he is likely to make to that site, because it doesn't only cost more to travel to such sites but also it takes more time to travel to forests in distant locations. TRIPTYPE has a relatively large positive sign suggesting that forest visit is a predominantly day trip recreation activity. The variable SITEHR has a negative sign and indicates that visitors travel more frequently to a forest site where they expect to spend less number of hours and vice versa. The fact that the variable SITEONLY has a positive sign, although not highly significant statistically, indicates that visitors often leave home for a single purpose trip to a forest site rather than combining such trips with visits to other activities. It is important to note that except for the variable SITEONLY, all other variables grouped under trip behaviour have not only the expected signs but are also statistically significant at 1 per cent level.

The variable INCOME appears to have a positive sign, significant at 1 per cent level, although this variable has been found non-significant in TC models, mostly applied to data from the US. Both variables AGE and SEX present positive signs which are significant at 1 per cent level. The effect of education, denoted by EDU, seems to be negative but non-significant. Zawacki et al. (2000, p. 17) state that 'income and education are perhaps too collinear to allow for independent estimation of the effect of education'. The results indicate that household size, HHOLDSIZE, has a positive but non-significant effect on trips to forest sites. Discretionary free time of visitors, FREEDAYS, is negatively related to number of forest visits per year and it is significant at 1 per cent level. As noted earlier, this might be explained by the fact that those who have a large amount of free time are likely to be those who live on low income (for example, unemployed and pensioners) and hence cannot afford to pay for frequent travels to visit forests located further away from their residential locations.

The coefficient of FORSIZE is positive and significant at 1 per cent, which means that the larger the total area of a forest site, the greater the number of times a visitor would travel to that site per year. Except for FORSIZE, the effect of each forest specific variable included in the model is not statistically significant. The existence of water features in a forested area seems to have a positive role in influencing a visitor to make more frequent trips to the area (Hynes et al.). The existence of trails for forest walk, denoted by the variable TRAILEN, has a very small and non-significant effect on trip behaviour. Similarly, population density within two hours drive from the forest areas appears to have a small but positive effect.

CONSUMER SURPLUS

CS is a widely used welfare measure in TC models. It is given as a difference between individual WTP and the actual expenditure on a good or service. This is obtained by calculating the integral above the average trip expenditure and below the estimated demand function. In applied models, the procedure most often used is to calculate CS per trip per person and this is given by the inverse of the coefficient of the TC variable (Creel and Loomis 1990; Yen and Adamowicz

1993). Martinez-Espineira and Amoako-Tuffour (2005, p. 20) state that, in TC models, only the coefficient on the TC variable is used to calculate welfare measures because expenses are mainly endogenous, a choice of the user. Forest visitors may also incur other expenses such as access fees but these usually constitute a small proportion of total TC of the visit.

Results given in Table 11.2 are used to calculate a welfare measure in terms of the CS that visitors derived from forest recreations. Thus, an inverse of the coefficient of out-of-pocket cost of travel, TRAVELCOST, yields a consumer surplus of £8.36 per person-trip for Model 1. The corresponding figure under Model 2 is given as £6.90. The difference in the amount of CS under Models 1 and 2 indicates that if the characteristics of avid users are not accounted for, then TC models would yield an overstated CS.

AGGREGATION OF CONSUMER SURPLUS

As long as the count data model is corrected for zero-truncation and endogenous stratification, Englin and Shonkwiler (1995) show that welfare measures for a given population could be calculated for the whole population. This means that, given the total number of forest visitors in Great Britain during a particular year, the per person-trip CS can be multiplied by the estimated number of population trips in a year to obtain the aggregate CS of access to forests throughout Great Britain.

Indeed, there has been a tradition of extrapolating the CS estimates into recreational use-benefits for the whole Forestry Commission estate (Willis and Bensen 1989; Scarpa 2003). It is important to note that such aggregations imply that the study sites covered are representative in terms of forest and visitor characteristics (Willis and Bensen 1989, p. 107). This study is based on data from 44 sites, a larger coverage than any other previous study as far as the author is aware, although this still may remain short of becoming a representative sample in some aspects of the data.

Aggregate CS is obtained by multiplying the person trip CS estimate by the total number of forest visits in Great Britain during the survey year. It was estimated that 252 million day visits from home were made to the woodland in Britain during 2002–3 (FC 2006). Applying estimated CS for Model 2, estimated aggregate CS would be £1739 million (£6.90 times 252 million) for all forest visits in Great Britain during the period.

COMPARISON WITH OTHER STUDIES

A useful summary and review of results from about 30 studies that have been undertaken on the value of recreation in UK woodland is provided by Christie et al. (2005, Table 6, p. 23). If values obtained from these studies, which were undertaken for the period 1970 to 1998, were indexed to account for inflation, then overall aggregate estimate of value for the UK would range between £9.1 million and £90 million.

Considering CS estimates obtained through different valuation methods (and again indexing for inflation), the range of values for three individual travel cost applications (the approach followed in this study) ranged from £0.10 to £4.00 per person-trip, which is a substantially smaller amount compared to the results of our study which ranged from £6.90 to £8.36. Eleven studies utilized a CVM and their CS value estimates ranged from £0.60 to £2.20 per person trip. This means that the CV technique did consistently yield even lower estimates. Average CS values obtained by three zonal travel cost studies ranged from £1.90 to £5.70, which means this approach yielded a relatively larger benefit estimates than the outcome of early applications of CV and TC methods to UK data.

After reviewing previous studies, Christie et al. (2006) went on to undertake their own estimates, applying a variety of standard valuation methods but here we concentrate on their findings from applying a TC model. Their study focused on generating CS estimates per person-trip separately for four groups of forest recreation users: cyclists, horse riders, walkers, and general visitors. For bird watchers, the estimated CS per person-trip was £7.90, which was the lowest figure they obtained when compared to the corresponding figures for the other groups. They identified relatively small variations in welfare benefits per person-trip for the other user groups: £14.97 for cyclists; £14.20 for horse-riders, £14.41 for walkers; and £14.99 for other forest visitors. Clearly, these values were substantially larger than those obtained from any other application of TCM to UK data, including our findings.

Christie et al. (2006) stated that the main reason for obtaining relatively high estimates was related to recent increases in the costs of travel (p. 7) and, additionally, they also indicated that there could be a number of methodological issues that may have contributed towards the higher values. First, they discarded survey data on participants whose return travel time was more than four hours in order to exclude those stayed over night. As they indicated, this would cause an upward bias. However, it actually seems that the main methodological element that may have caused overestimation of their estimate of CS per person-trip was the choice of econometric estimation procedure. To begin with, they accounted for neither zero-truncation nor endogenous stratification of the data. Most importantly, they employed a simple Poisson Count Data Model which does not correct for possible over-dispersion of the data. They reported that they were not successful in implementing a NEGBIN model on the ground that the computation process for this model did not lead to convergence.

In summary, a fairly sophisticated econometric procedure is applied in our study, controlling for socio-economic characteristics of visitors as well as unique features specific to each site. Consequently, robust econometric results were obtained, the estimates falling within reasonable limits of estimates obtained by previous TC models applied to UK data.

Conclusion

The CS estimates from TC-based studies may differ due to a variety of reasons. Some of these could be attributed to unresolved conceptual issues or definitional ambiguities, such as opportunity cost of travel time or substitution effects among adjacent sites. Other sources of differences in non-market benefit estimates from TC models could be attributed to econometric estimation procedures, specifically, the selection of a suitable econometric model that would best fit the data. By drawing attention to deficiencies in the literature, particularly in regard to econometric model selection and specification, this study has made a modest contribution to the UK literature on application of the TC model to valuations of forest recreation benefits.

As noted earlier, the main shortcomings of the TCM models are that they are based on on-site sampling and this involves inherent bias in two respects: the information refers to only those who visited the site during the survey and even then a good proportion of those could be enthusiastic and frequent forest visitors. However, there have been methodological advances in count data modelling to correct for the inherent weakness of TC approach. One can either undertake an off-site survey and then base the econometric estimation on a data set created by pooling on-site and off-site surveys or alternatively it is possible to employ a sophisticated econometric procedure to correct for sample bias. In this study, the second alternative was implemented; a standard NEGBIN model was adjusted for zero-truncation and endogenous stratification. Consequently, the estimated coefficients and the calculated per person-trip CS obtained was just about the average for a relatively large number of previous TC studies in the UK. This indicates that the econometric results of this study may have minimized the over- or under-statement of non-market benefits from forest recreation, which have frequently been observed in applications of TC models to UK data.

The main contribution of this study is the application of zero-truncated and endogenously stratified NEGBIN to UK data for the first time, as far as the author of this paper is aware. It has drawn attention to the need to improve the precision with which non-market benefits of forest recreation are measured. As noted in the introduction, there has been a growing demand by policy makers for evidence-based decisions regarding the use of public money to meet the growing demand for forest recreation. Thus, researchers need to be concerned with the rigour with which non-market benefits are estimated, given the potential use of such estimates in vital public decisions to allocate scarce resources.

It proves useful to bear in mind that, no matter how accurately they are measured, TC model-based benefits provide only a fraction of total benefits that society derives from forest recreation. The aggregate CS estimated in this study represents non-market economic value (benefit) but it excludes market-based benefits which can be estimated using standard multiplier analysis. For instance, the market-based benefits to society, for example, visitor expenditure on local economies of the forest site, are not included in benefits obtained from TC models. The estimation of total economic benefit, which combines market and non-market benefits, has been left for future research. However, it has to be noted that even the total economic benefit (or total economic value) represents a conservative estimate of societal gains from forest recreation because of the existence of non-economic benefits from forests such as bequest value of forests as part of the natural environment that need to be preserved and passed on to the next generation.

REFERENCES

Bhat, M.G. (2003), 'Application of Non-Market Valuation to the Florida Keys Marine Reserve Management', *Journal of Environmental Management*, Vol. 67, No. 4, pp. 315–25.

Bin, O., C.E. Landry, C. Ellis, and H. Vogelsong (2005), 'Some Consumer Surplus Estimates For North Carolina Beaches', *Marine Resource Economics*, Vol. 20, No. 2, pp. 145–61.

Bowker, J.M., D.B. K. English, and J.A. Donovan (1996), 'Toward a Value for Guided Rafting on Southern Rivers', *Journal of Agricultural and Applied Economics*, Vol. 28, No. 2, pp. 423–32.

Christie, M., N. Hanley, T. Hyde, B. Garrod, K. Swales, N. Lyons, I. Keirle1, and A. Bergmann (2005), 'Valuing Forest Recreation Activities: Phase 1 Report to the Forestry Commission, mimeo, London.

Christie, M., N. Hanley, B. Garrod, T. Hyde, N. Lyons, A. Bergmann, and S. Hynes (2006), 'Valuing Forest Recreation Activities: Final Phase 2 Report to the Forestry Commission.

Creel, M. and J.B. Loomis (1990), 'Theoretical and Empirical Advantages of Truncated Count Data Estimators for Analysis of Deer Hunting in California', *American Journal Of Agricultural Economics*, Vol. 72, pp. 434–41.

Dennis, D.F. (1998), 'Analyzing Public Inputs to Multiple Objective Decisions on National Forests Using Conjoint Analysis', *Forest Science* Vol. 44, pp. 421–9.

Englin, J.E., T.P. Holmes, and E.O. Sills (2003), 'Estimating Forest Recreation Demand Using Count Data Models, in Forests in a Market Economy, Sills, E.O. (ed.), Dordrecht, The Netherlands: Kluwer Academic Publishers.

FC (Forestry Commission) (2006), Forestry Statistics 2006, Edinburgh: Forestry Commission.

Hill, G., P. Courtney, R. Burton, J. Potts, P. Shannon, N. Hanley, C. Spash, J. Degroote, D. Macmillan, and A. Gelan, 'Forests' Role in Tourism: A Report for the Forestry Group (Economics and Statistics) of The Forestry Commission, *http://www.Forestry.Gov.Uk/PdffTourismfinal.Pdff $File/Tourismfinal.Pdf.* (accessed on 30/10/2006).

Hynes, S., N. Hanley, and C. O'Donoghue, 'Measuring the Opportunity Cost of Time in Recreation Demand Modeling: An Application to Whitewater Kayaking in Ireland', *http://www.economics.stir.ac.uk/staff/ hanley/sitechoice4.pdf* (accessed on June 5, 2007).

Leeworthy, V.R. and J.M. Bowker (1997), 'Non-market Economic User Values Of The Florida Keys/Key West: Technical Report, National Oceanic and Atmospheric Administration, Strategic Environmental Assessments Division,', mimeo, Silver Spring, MD.

Liston-Heyes, C. and A. Heyes (1999), 'Recreational Benefits from the Dartmoor National Park', *Journal of Environmental Management*, Vol. 55, No. 2, pp. 69–80.

Loomis, J. (2003), 'Travel Cost Demand Model Based River Recreation Benefit Estimates with On-Site and Household Surveys: Comparative Results and a Correction Procedure', *Water Resources Research*, Vol. 39, No. 4, pp. 1–15.

Martinez-Espineira, Roberto and Amoako-Tuffour, Joe, 'Recreation Demand Analysis under Truncation, Over-dispersion, and Endogenous Stratification: An Application to Gros Morne, National Park Economics, St. Francis Xavier University, *http://129.3.20.41/Eps/Em/Papers/0511/0511007.Pdf* [accessed November 1, 2006]

NAO (1986), National Audit Office, Review of Forestry Commission Objectives and Achievements: A Report by the Comptroller and Auditor General, National Audit Office, London: HMSO.

Scarpa, R. (2003), The Recreational Value of Woodlands, Edinburgh: Forestry Commission.

Shrestha, K.R., A.F. Seidl, and A.S. Moraes (2002), 'Value of Recreational Fishing in the Brazilian Pantanal: A Travel Cost Analysis Using Count Data Models, *Ecological Economics*, Vol. 42, Nos 1–2, pp. 289–99.

Sohngen, B., F. Lichtkoppler, and M. Bielen (2000), 'The Value of Day Trips to Lake Erie Beaches', Technical Report TB-039, Ohio Sea Grant Extension, Columbus OH.

Ward, F. A. and J.B. Loomis (1986), 'The Travel Cost Demand Model as an Environmental Policy Assessment Tool: A Review of Literature', *Western Journal of Agricultural Economics*, Vol. 11, No. 2, pp. 164–78.

Willis, Ken (2003), 'The Social and Environmental Benefits of Forests: Report to the Forestry Commission', mimeo, Edinburgh: Forestry Commission.

Willis, K., G. Garrod, and R. Scarpa, D. Macmillan, and I. Bateman, (2000), 'Non-Market Benefits of Forestry: A Report to the Forestry Commission', mimeo, Forestry Commission: Edinburgh.

Willis, K.G. and G.D. Garrod (1992), 'Amenity Value of Forests in Great Britain and Its Impact on the Internal Rate of Return From Forestry', *Forestry*, Vol. 65, No. 3, pp. 331–46.

――― (1993), 'Valuing Landscape: A Contingent Valuation Approach', *Journal of Environmental Management*, Vol. 37, pp. 1–22.

Willis, K.G. and G.D. Garrod (1991), 'An Individual Travel Cost Method of Evaluating Forest Recreation', *Journal of Agricultural Economics*, Vol. 42, pp. 33–42.

Willis, K.G., G.D. Garrod, J.F. Benson, and M. Carter (1996), 'Benefits and Costs of the Wildlife Enhancement Scheme: A Case Study of the Pevensey Levels', *Journal of Environmental Planning and Management*, Vol. 39, pp. 387–401.

Willis, K.G. and J.F. Benson (1989), 'Values of User Benefits of Forest Recreation: Some Further Site Surveys: A Report to the Forestry Commission', mimeo, Department of Town and County Planning. University of Newcastle Upon Tyne: New Castle.

Yen, S.T. and W.L. Adamowicz (1993), 'Statistical Properties of Welfare Measures from Count-Data Models of Recreation Demand', *Review of Agricultural Economics*, Vol. 15, pp. 203–15.

Zawacki, William T., Allan Marsinko, and J.M. Bowker (2000), 'A Travel Cost Analysis of Non-consumptive Wildlife-Associated Recreation in the United States', *Forest Science*, Vol. 46, No. 4, pp. 496–506.

Zinkhan, F.C., T.P. Holmes, and D.E. Mercer (1997), 'Conjoint Analysis: A Preference-Based Approach for the Accounting of Multiple Benefits in Southern Forest Management', *Southern Journal of Applied Forestry*, Vol. 21, pp. 180–6.

12

Tourism, Local Livelihood, and Conservation

A Case Study in Indian Sundarbans

INDRILA GUHA AND SANTADAS GHOSH

INTRODUCTION

Studies have already identified that the dual goal of poverty eradication and conservation can be effectively achieved by tourism which is nature-based and 'pro-poor'. Protected national parks constitute a significant market for tourism based on natural resources and local culture. Tourism needs to be organized in ways that enable local people to have better access to tourists so that they can augment their livelihood through employment and small enterprise development (Goodwin 2002; Ashley 2002). In recognition of this, the Pro-Poor Tourism (PPT) guidelines and the idea of 'responsible tourism' have been developed [www.propoortourism.org.uk]. Market-driven private commercial enterprises may not ensure adherence to such principles by themselves. As a result, the effectiveness of PPT strategies depends enormously on the local conservation authority (Ashley et al. 1999). Literature identifies tourism as a very effective poverty eradication

This study is the partial outcome of a research project funded by the South Asian Network for Development and Environmental Economics (SANDEE) and hosted by the Global Change Programme, Jadavpur University. We benefited immensely through various Bi-annual R&T Workshops of SANDEE. We are thankful to Enamul Haque, Priya Shyamsundar, Jeffrey Vincent, Kanchan Chopra, Karl-Goran Maler, M.N. Murty, Gopal Kadekodi, Puran Mongia, Subhrendu Pattanayek, E. Somanathan, K. Gunnard, and S. Madheswaran for their valuable comments and suggestions at various stages of this study. We also express our sincere gratitude to Joyashree Roy for helping us to locate this project in the Global Change Programme, Jadavpur University, and for providing institutional support throughout the study period.

tool in a remote rural set-up. It supports livelihood diversification, which is particularly relevant to remote areas. Also, it is more labour-intensive compared to other industries, can grow with unskilled labour, and has low entry barriers (Holland et al. 2003; Elliott 2001).

Sundarban Tiger Reserve (STR), in the eastern part of India, is a part of the largest inter-tidal area in the world covered by mangrove forest. It renders important ecological services to a vast region in South Asia. The STR is a pack of forest-islands with no human habitation within it. Rivers or water channels separate all the surrounding island villages. The residents in these villages are far from the mainland. The forest provides livelihood opportunities for the local poor who traditionally depend on the forest for its products and fish in its water channels. The island villages have come up within the last hundred years. They have been settled by migrants and there seems to be no indigenously developed local institution to ensure the sustainable use of the forest. There is wide recognition that the forest quality has deteriorated significantly over the last few decades and that the man–forest conflict is intense. Also, the forest is home to the endangered Royal Bengal Tiger whose man-eating propensity is historically high. This also renders the man–animal conflict in the reserve as being a problem of special dimension.[1] As part of the nationwide conservation programme, 'Project Tiger', and situated within a declared World Heritage Site (1985), the management of STR is being attempted by effectively blocking human intrusion. This is resulting in conflicts between government-sponsored conservation efforts and the livelihood opportunities of locals. For resolving such conflicts, the authority (Department of Forest) has recently initiated some alternative income generating schemes to reduce forest dependence of the locals. However, a carefully devised tourism promotion initiative is conspicuous by its absence.

An effective conservation strategy for a forest must turn the local poor from intruders into its keepers by making them stakeholders in the earning opportunities that the conservation provides. The STR is a great tourist attraction and has seen a surge in the number of visitors in recent years. Tourism may act as a vehicle for conservation by providing new livelihood opportunities in Sundarbans' remote

---

[1]MOEF (Govt. of India) *http://projecttiger.nic.in/sundarbans.htm;*
WWF: *http://www.wwfindia.org/about_wwf/what_we_do/tiger_wildlife/our_work/tiger_conservation/sunderbans/*

village economy. Empirical study on this aspect is needed for informed decision making by the authority and integrating local stakeholders into the conservation programme in a sustainable way. This study investigates the possible effects of tourism on local livelihood and thereby intends to fill up the crucial information gap.

BACKGROUND

The economy of the island villages of Sundarbans is characterized by remoteness, and the absence of electricity, power-driven industry, or any nearby urban centre to sell local products. Villagers have few occupational choices other than agriculture and fishing. However, these bottlenecks can be turned into assets for developing nature-based tourism, and bringing in the local poor as stakeholders. Organized and regulated tourism in the Indian Sundarbans started since the mid-1980s, with the inception of 'Project Tiger'. For tourists, permits (priced) to enter the forest became mandatory. Permits are issued at four different places but there is a single entry point to STR. The village Pakhiralay, overlooking the forest and at one corner of a larger inhabited island, hosts the Forest Range Office where all permits are to be produced before entering the forest. Consequently, it became the village receiving visitors to STR. A Sundarban tour is essentially a cruise through the water channels within the forest with halts at a handful of watch-towers on riverbanks. There are package tours originating elsewhere and touching Pakhiralay for permits only. The tourists spend the nights on launches (large watercrafts) where food is also cooked and served on board. Such package tours are adopted by approximately 70 per cent of the visitors. There is little scope for local villagers to trade with such visitors. But tourists in smaller groups also come and stay in Pakhiralay and hire boats (smaller watercraft) locally for a day-long cruise into the forest. With an increase in demand in this segment (estimated at 30 per cent), many tourist lodges have came up in the village in the last few years with the attendant business opportunities. Villagers have spontaneously availed these opportunities as owners of tourist lodges/boats, boat drivers, forest guides, cooks, drinking water suppliers, owners of small shops and telephone booths, and so on.

Focus group discussions revealed that people directly exploiting the forest are mostly landless and poor who live on the riverbanks across the forest. Almost all such households now come under one of the many Eco-Development Committees (EDC) of the STR authority. These EDCs are being provided with finances under various schemes of alternative income generation. This study intends to find out whether new earning opportunities thrown open by tourism are significantly augmenting livelihood opportunities for local households. In doing so, it restricts the definition of 'local households' to 'the river-side households in Pakhiralay'—this largely coincides with the EDC membership. The 'Greater Sundarbans' is a vast populated area and most of its residents have no direct link with the forest. So, a broader definition for 'local' would dilute the possible policy relevance of the study findings.

Also, sustainable and responsible tourism practices call for ensuring the maintenance of cultural assets along with natural assets of a community. Existing literature on responsible tourism identifies negative social impacts of tourism and calls for assessing such impacts through qualitative responses from local stakeholders (Spenceley 2001).

STUDY OBJECTIVE

With this background, the study is intended to achieve the following two specific objectives:

1. To empirically measure the extent to which tourism is augmenting local livelihood in the village Pakhiralay, as revealed by households' expenditure pattern.
2. To find out local villagers' perception regarding tourism's possible positive and negative social effects by eliciting qualitative responses from them.

EXISTING LITERATURE

The issue of protecting environment by carefully encouraging tourism that augments the livelihood opportunities of the local poor has been examined in several policy-oriented studies (Alpizar 2002; Andersson 2004; Sills 1998). The specific aspect of tourism–poverty interaction is dealt with more elaborately in the PPT literature. It identifies several reasons why tourism may be particularly effective in reducing local poverty and peoples' forest dependence. The following advantages of tourism as a vehicle for poverty eradication and conservation have been identified (Elliott 2001, Ashley et al. 1999, 2002):

• High income-elasticity, and therefore offers a relatively rapidly growing market.

- More labour-intensive than other sectors providing diverse employment opportunities for people with a wide range of skills, as well as for the unskilled.
- Not import-intensive, making it particularly attractive to developing countries/backward areas.
- Higher potential for linkage with local enterprises because customers come to the destination.
- Low entry barriers.
- Tourism products can be built on natural resources and culture, making them productive assets.
- Helps locals to enjoy better infrastructure (water, health, and communication), security (law and order), and better information about the outside world.
- A wise tourism policy contributes to the protection of the assets such as wildlife and plant diversity.

Case studies have found that in spite of the above mentioned potentials, tourism is often characterized by a myopic private sector, limited involvement of local communities, and lack of market access (Spenceley 2001). Also, impacts of different tourism segments and types of tourism on the local poor are different. These call for strategies/guidelines for PPT and authority interventions at the local level (Ashley 2002; Ashley et al. 1999, 2002).

The existing literature points out that there can be negative social effects of tourism on local communities in remote rural set-ups in terms of cultural shocks from outsiders (Spenceley 2001). However, studies also indicated that generally most effects were perceived as positive by local residents. A study from Equador finds that the encounter with the tourist is mostly seen as a positive experience of cultural exchange by the local stakeholders (Wunder 1999, 2000). Attitudes towards tourism in local communities adjacent to national parks in Indonesia and Zimbabwe show significant optimism in response to similar kinds of questions (Goodwin 2002).

Studies point out that in areas with significant seasonal variation in the number of visitors, most of the people treat tourism-related income as an 'additional' revenue to subsistence farming (Saville 2001). Tourism can boost local arts, culture, and traditional medicine. These in turn may boost up tourism. Such tourism may change local people's attitudes towards preservation and conservation of flora and fauna (Kulkarni et al. 2002).

Literature shows that different segments of tourists have variable implications for income generation for the local residents (Holland et al. 2003). Tourism in 'all inclusive' packages is of the 'enclave form', where those wishing to sell to tourists are often reduced to hawking at the enclave entry and exit points (Goodwin 2002). A study of Himalayan trails in Nepal with lack of roads and transport facilities shows that backpackers eating and staying in local hotels generate more money for locals than pre-paid organized treks (Saville 2001). A study from Brazil revealed that low-income tourism in a village can generate sizeable income for local entrepreneurs (Wunder 2003).

In the Indian context, more than one illuminating studies are available for Keoladeo National Park, identifying and estimating the impact of tourists on incomes of different related service providers (Chopra 2004; Goodwin 2002; Goodwin et al. 1997). In another study on Pench National Park, similar identification of local beneficiaries and quantitative estimation of their annual tourism earnings is available (Kulkarni et al. 2002).

There have been recent studies on the environmental and socio-economic impacts of shrimp farming in the Indian Sundarbans (Chopra and Kumar 2005), which is perhaps the most important commercial activity in this area. There are also studies on the valuation of timber and non-timber forest products (Santhakumar et al. 2005). However, the literature does not provide any empirical study on tourism and/or its impact on the local economy in Indian Sundarbans.

STUDY AREA AND DATA

The Sundarbans is a continuous mangrove delta region covering both India and Bangladesh. Approximately one-third of its area falls in India. Members of landless and marginal households (HHs) in the fringe villages of STR often venture into the forest to catch fish and crab in the creeks and also to collect firewood and honey with permits for a limited period. The spatial distribution of population within the surrounding islands is closely linked with their occupational distribution. Landless and marginal HHs, which are often directly dependent on forest and rivers, are concentrated on the river-banks bordering the forest. The landed HHs are mostly placed in the interiors or towards the mainland.

Tourist arrival in Sundarbans is concentrated almost entirely during the winter months as the absence of electricity, adverse climatic conditions, and choppy rivers discourage tourists during other parts of the year. During this peak tourist season, local villagers participate in various capacities as service providers which have already been mentioned. Extreme seasonal variation in tourist arrival provides little opportunity for a village HH to depend entirely on tourism as a year-long occupation. Every participating HH has some other earning sources availed by all or some of its members. All tourist lodges and related business establishments in Pakhiralay, some purely seasonally operated, are concentrated along a 500 metre stretch of road on the riverfront. It is found that tourism participants almost invariably belong to the riverside population. Interestingly, this population also includes the directly forest dependents.

In this set-up, a study on the role of tourism in the context of conservation calls for an empirical study on the livelihood of riverside

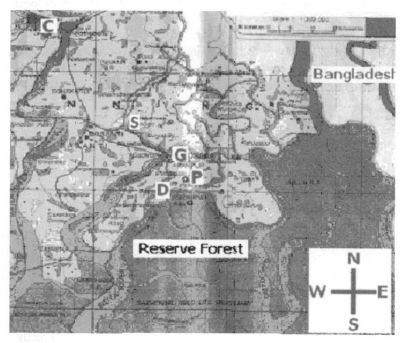

P–study village (Pakhiralay)   D–control village (Dulki)
G–wholesale market (Gosaba)   S–nearest road link with mainland
C–nearest railway station and town (Canning)

Map 12.1:   Sundarbans, West Bengal

Note: Map not to scale.
Source: *http://s24pgs.gov.in/index.htm*

households. To filter out the direct and indirect effects of tourism on livelihood, a control village with similar geographical and socio-economic characteristics needs to be incorporated. In this study, the 'study village' (Pakhiralay) and the 'control village' (Dulki) are parts of the same bigger island. Both are placed across the forest and on the bank of the same river, with very similar soil conditions and other socio-economic features. Both are located in the eastern fringe (forest side) of the island while the nearest wholesale market (Gosaba) is on the western tip.

Together, the market and the two villages form a triangle by their geographical location. One important feature is that while each of these villages is linked to Gosaba market by village roads, there is no proper road link between them. As a result, there is little economic interaction between the study village and the control village. This renders percolation of tourism money from study village to control village practically impossible.

The contribution of tourism money in augmenting the livelihood of local HHs is estimated by a primary survey of HH consumption expenditure. The sample HHs have been selected by stratified

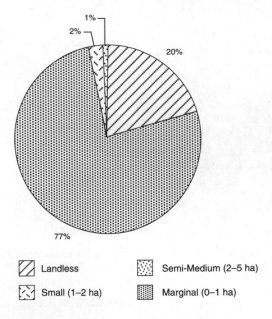

Fig. 12.1:   Distribution of Landholding

random sampling from the study and control villages. The stratification has been done according to HH's landholding status. This study has generated lists of HHs located within 500 metres from the riverfront for both the villages along with their latest landholding status. The sampling frame consisted of 273 HHs in the study village and 193 HHs in the control village. It was found that these riverside HHs are mostly 'landless' or 'marginal'.[2] No 'medium' or 'large' landholding is found in the selected area. The proportion landless:marginal:small:semi-medium is found to be 70:194:6:3 in the study village while it is 25:162:5:1 in the control village. The landholding distribution of the survey population is found to be heavily skewed towards landless and marginal farmers.

From each of the four landholding strata, approximately 18 per cent of the HHs were selected through equal probability sampling. This resulted in a data size of 87 HHs (48 in study village and 39 in control village).[3] Each of the sample HHs was visited once during off-season for tourism (August–September 2005) and again once during peak season (February–March 2006). The expenditure data were collected in two rounds, while other HH level variables were supposed to be time invariants within a span of less than six months and were recorded in the first round alone. The HH survey questionnaire was appended with a module with queries relating to respondents' perception on possible social and cultural effects of tourism in the village. This appended module was intended only for the respondent HHs in the study village in the second round (peak season).

Methodology

It can be generally stated that the livelihood opportunities of a HH depend on: (i) physical capital in its possession; (ii) quality of natural capital that it has access to (iii) public capital (infrastructure); and (iv) its human capital. In the Indian Sundarbans, apart from private capital (mostly cultivable land), the natural capital is the forest and the river. Direct conservation efforts are gradually lowering locals' access to this capital. Also, unsustainable fishing and forest

[2]Marginal (0–1 ha); small (1–2 ha); semi-medium (2–5 ha); medium (5–10 ha); large (> 10 ha). Source: Ministry of Agriculture, Government of India.

[3]Number of sample HHs from control village is marginally more than 18 per cent, for rounding off numbers in each stratum.

exploitation is affecting its quality. The geographical isolation of the region and poor accessibility has resulted in insufficient infrastructure (public capital).

Local private physical capital formation, apart from the bottleneck of infrastructure, is crippled by a lack of local surplus. Tourism can open up new vistas for livelihood opportunities for a community apart from traditional production activities. It can make hidden social and natural capital marketable and productive. Social culture and the existence value of the forest can turn into marketable productive assets.

The 'direct' local impact of tourism money can be manifested by additional consumption expenditure by the HHs participating in tourism. However, the total impact on local HHs is expected to be larger as some of the additional expenditure by participating HHs can find its way to other non-participating HHs through intra-village transactions. The extent of these secondary benefits crucially depends upon the extent to which villagers spend their additional income on products produced within the village. In this context, existing studies found significant 'leakages' in tourism money which escapes from the local communities. Estimates show that 50 per cent to 90 per cent of tourists' spending usually leaks out of communities closest to nature attractions (Buchsbaum 2004, Goodwin 2002, Holland et al. 2003). Estimating such leakages is difficult, but it seems that a large part of tourism money received by the villagers is spent on consumer durables such as radios, solar energy cells, bicycles etc. which are imported from outside the village.

Literature also identifies that a portion of additional income from tourism is used to finance children's education. This enhances the human capital of village HHs and can open up possibilities for new occupations within the region or migrating outside. It may also provide the basis of informed decision making vis-à-vis conservation and sustainable use of natural resources on the part of the local stakeholders and embolden them to enter into a dialogue with authorities and outsiders. This exemplifies the complex character of social changes activated by the cash inflow from tourism (Wunder 1999, 2000). It calls for a deeper analysis of the expenditure pattern of participating vis-à-vis non-participating HHs. A significantly higher spending on education by participating HHs may be indicative of tourism acting as a trigger for an endogenous progress in human capital formation. In that case, tourism can be conceived as the vehicle for a self-sustainable conservation programme.

The first objective of this study has been to empirically measure the livelihood-augmenting impact of tourism at the local level. In doing so, the study investigates both the 'direct' and 'indirect' effects of tourism money. As already stated, the sample HHs come from a population which is mostly landless or marginal. Focus group discussions suggest nil or insignificant savings accrue to these HHs on a yearly basis. In this set-up, the livelihood status of a HH is assumed to be reflected by its monthly per capita 'expenditure' in monetary terms. However, the 'expenditure' figures are arrived at by taking into account the consumption of self-produced commodities which are valued at prevailing market prices. For a deeper insight into the affordability and attitude of different HHs, the expenditures are classified as 'food expenses', 'non-food expenses', and 'expenses on education of children'. While the first two may be regarded as HH consumption, the third may be considered as investment for developing human capital.

The 'direct' effect of tourism on livelihood is examined by differences in welfare indicators between participant and non-participant HHs in the study village (*Pakhiralay*). The essence of incorporating a control village is to examine the presence of any significant 'indirect' or 'trickle-down' effect of tourism money on expenditures of non-participating HHs, possibly accrued through intra-village transactions. The difference across groups and across villages has been examined through some mean comparisons in welfare-indicators as well as other HH-level variables.

Identification of determinants of HH welfare has been carried out through regression analysis. Information on expenditures was collected for each HH in two rounds coinciding with off and peak season of tourism in the study area. This was done to check whether the tourism money earned in the peak season by some of the HHs was retained to finance their off-season expenditures. Per capita monthly monetary expenditure has been regressed on HH level variables including landholding as well as dummy variables for tourism-participation, season, and village. This has been done through an OLS regression with robust standard errors.

Additionally, expenditures on 'food', on 'non-food items', and 'per child monthly monetary expenditure on education' have also been regressed on relevant HH level variables. In recognition of the possible contemporaneous correlation in error terms for the three equations, the estimation has been carried out as a system of Seemingly Unrelated

Regression Equations (SURE). To check the robustness of the estimates, regressions have been run for the pooled data on two villages as well as on the data from the study village alone.

The second objective of this study has been to find out the local villagers' perception regarding the possible positive and negative social impacts of tourism. For this, responses on qualitative aspects of tourism on village life were elicited through 'yes', 'no', and 'indifferent' choices. The responses were elicited from the study village and only in the peak tourist season (winter) as the villagers' perception is expected to be more focused at this time. Since the sample HHs were selected randomly with landholding-based stratification, the aggregate analysis of these responses is expected to represent the village psyche on tourism. Aggregative descriptive statistics have been used in this part of the study.

EMPIRICAL RESULTS

Profile of survey households

The survey data show that the riverside population in the study area (local HHs) is indeed worse off as compared to their rural counterparts in the state (West Bengal). Distribution of HHs across monthly per capita expenditure (MPCE) class shows the median value to be Rs 500 for rural West Bengal in 2004–5[4] while that for the sample HHs, it is found to be Rs 433 in 2005–6 (Appendix A12.1.1).

To bring out the implication of tourism for conservation, the study compares HHs that have been classified as 'tourism-participants', 'forest-dependants' and 'engaged in other economic activities'. Human intrusion into the forest takes place primarily in two forms. First, the honey and wood collectors (though wood collection has recently been officially banned) and the fish and crab catchers intrude into the forest's interiors regularly. Second, villagers venture into prawn-fry collection in the rivers in a very crude way causing much damage to the delicate ecosystem in the process. Both these operations are perceived to involve real danger due to tigers in the forest and crocodiles in the rivers.[5] It can be perceived that these poor HHs on the riverside directly depend on the forest as a last resort for their livelihood.

[4]Level and Pattern of Consumer Expenditure, 2004–05; NSS 61st Round: NSSO; Ministry of Statistics and Programme Implementation, GoI. *http://mospi.nic.in/mospi_nsso_rept_pubn.htm* (Table 1R).

[5]Man–animal conflict in Sundarbans: *http://projecttiger.nic.in/sundarbans.htm*

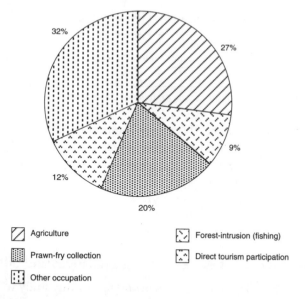

▨ Agriculture	⊠ Forest-intrusion (fishing)
▦ Prawn-fry collection	⊡ Direct tourism participation
⊞ Other occupation	

Fig. 12.2:   Occupational distribution of working adults (primary occupation)

Being mostly landless or marginal, few HHs depend on agriculture alone. Often different working members of one HH are engaged in different occupations. The occupational distribution can, therefore be analysed at an individual level rather than at the HH level. This distribution for the study village[6] shows that agriculture is the single occupation employing maximum number in terms of working population in the study area. 'Other occupation' represents a composite of many occupations at the village level[7] including daily labourers who get some seasonal employment in agriculture. Thus, the share of agriculture shown here is an underestimate. It is notable that there is no industrial workforce in the area. This is due to the absence of any power driven industry.

In a few cases, one HH can be identified with a single occupation since all its working members are engaged in that occupation. Table

[6]Since tourism is an occupation only in the study village, Figure 13.2 is obtained for the study village only.

[7]'Other occupations' include agricultural labourers, traders, cycle van pullers (the only mode of transport on the island), salaried employees (public and private), and other local skilled workers and professionals (artisans, priests, private tutors, quacks, masons, and carpenters).

Table 12.1:   Distribution of sample HHs with a single source of livelihood*

Occupation	Agriculture	Forest intrusion	Prawn-fry collection	Tourism participation	Other occupations	Total
No. of HHs	7	3	2	0	13	25

Notes: *Out of 87 HHs in the sample (both study and control villages).

12.1 describes 25 such HHs out of the total sample of 87. It shows that relatively greater number of them sustain only with agriculture or 'other occupation'. Forest-intrusion or Prawn-fry collection sustain a much smaller number. This can be indicative of dwindling forest quality as a livelihood option. Significantly, none of the sample HHs sustains itself only through tourism-jobs. This is obvious as tourism in Sundarbans, as of now, is restricted to the winter months only.

A tale of two villages—examining similarities and differences

The control village has been incorporated in this study to find out differences at the village level resulting from the presence and absence of tourism. For any 'indirect' or 'trickle-down' effect of tourism money, welfare indicators (expenditures) of non-participating HHs have to be significantly different across the two villages. A comparison in mean values for similarities in HH level characteristics and differences in HH level expenditures has been carried out. For a statistically valid inference, the control village needs to be geographically and socio-economically similar to the study village except for the existence of tourism.

Table 12.2:   Village level information from secondary sources

	Study Village	Control Village
Area (ha)	479.49	419.39
Number of households	772	566
Population	3871	2710
Sex ratio (M/F)	1.018	1.016

Source: Directorate of Census Operations, West Bengal (2003).

In the rural island villages of Sundarbans, villages are very similar in their infrastructure, location vis-à-vis the mainland, and their socio-economic characteristics when they are part of the same island. Also, prices in local markets within the same island are comparable while they vary across islands due to differential transport cost vis-à-vis the mainland. In this study, the study and control village are both parts of the same island, on the banks of the same river. However, it was found that there is a larger concentration of landless and marginal farmers in the study village.[8] Table 12.2 is based on secondary source[9] which shows a larger number of HHs in the study village, but an insignificant difference in the total land area. In spite of this difference, the control village has been selected due to its similarity with the study village in terms of infrastructure, market prices and geography, apart from advantages in terms of survey logistics. It is perceivable that this difference between the two villages narrows down in the study as it includes only a subset of village population (riverside residents) into its sampling frame, which is expected to show homogeneity in HH level variables across villages.

It is important to check whether some crucial HH level variables differ across the two villages significantly before a statistically valid inference can be made from regression analysis. Since the sample sizes are not large, and the tourism participation is entirely concentrated in one of the villages, an aggregative comparison at the village level (from the survey data) may be misleading. Instead, for village level comparison, HHs have been divided into three categories: (i) HHs participating in tourism-related jobs, (ii) HHs having direct forest-dependence, and (iii) HHs which are neither tourism-participants nor directly forest-dependents. While the first category can only be found in the study village, comparison has been carried out for the other two categories across the villages.

The null hypothesis of equality of mean values across two villages is tested for two different levels of significance. The acceptance/rejection status of null hypothesis for some crucial HH level variables is shown in Table 12.3. The results show that the characteristics of the riverside HHs of these two categories, in terms of family size and endowments, do not significantly vary across the two villages.

Table 12.4 describes the expenditure pattern across the two villages, taking all categories of HHs together. It is found that mean values of per capita monthly expenditure on food, non-food items, and per child monthly expenditure on education are marginally more in the study village, though the differences are not statistically

[8]Mentioned in the third section: the proportion of landless:marginal:small:semi-medium is 70:194:6:3 in the study village while it is 25:162:5:1 in the control village.
[9]*Source*: Directorate of Census Operations, West Bengal (2003).

Table 12.3:   Test for equality of mean values across villages

HH level variables	Directly forest-dependent HHs (and not tourism-participants)		HHs neither tourism-participants nor forest-dependent	
	Level of significance		Level of significance	
	5%	1%	5%	1%
Avg. family size	Accepted		Rejected	Accepted
% of literacy among adults	Rejected	Accepted	Accepted	
% of adults completing primary education	Accepted		Accepted	
Average HH landholding (*katha*)	Accepted		Accepted	
Average per capita landholding (*katha*)	Accepted		Accepted	
Proportion of HH's owning livestock	Rejected	Accepted	Accepted	

Notes: Null hypothesis: mean values are equal across study and control village.
Katha is the smallest local unit of landholding; 1 hectare = 149 Katha (approximately).

Table 12.4:   Variation in average per capita expenditure across two villages

HH level Per capita Expenditures (Item)	Control Village Mean (SD) (Rs.)	Study Village Mean (SD) (Rs.)
Per capita monthly expenditure on food (Rs.)	266.69 (77.01)	271.22 (115.65)*
Per capita monthly expenditure on non-food items (Rs.)	162.86 (61.96)	169.14 (110.76)*
Per child monthly expenditure on education (Rs.)	40.89 (52.39)	62.65 (102.80)*

Notes: Sample size: control village: 39 HHs, two rounds; study village 48 HHs (two rounds). *Significant difference in SD—Standard deviation across villages.

significant. However, the SD are markedly higher in the study village. This greater heterogeneity in the study village calls for a closer examination. It might be indicative of existence of subgroups of HHs with significantly different expenditure patterns. It is plausible that the heterogeneity is caused by tourism participation, which is absent in the control village. It is also conceivable that as tourism participation is not yet widespread in the study village, the weight of participating households is not large enough to reveal a significant inter-village difference in the mean values. This calls for a deeper investigation of characteristics and expenditure patterns of HHs across subgroups.

Understanding tourism participants and the rest: data from study village

Since tourism-related activities and establishments are concentrated in a small part of the study village and all participants are local residents, it was possible to enumerate all of them according to their mode of participation. The number of persons engaged in services/ trades catering exclusively to tourism is found to be 77 in 2005–06 which is approximately 8.2 per cent of the adult population.[10] In addition, there are trades (grocery shops, telephone booths, local ferry service) which do not cater exclusively to tourists, but gain significantly by their arrival. Approximately 6.9 per cent of local adult population is engaged in these trades. However, as the latter set of locals operates their trades throughout the year catering to the villagers, this study excludes them from the category of 'direct tourism participants'. Therefore, the study findings regarding tourism's livelihood contribution is somewhat an underestimation. Inclusion of these additional gains from tourism will certainly increase the study estimates.

As tourism is absent in the control village, a comparison between 'direct tourism participants' and the 'rest' can be done with the sample HHs from the study village only. Table 12.4 has already indicated the possible existence of subgroups of households in the study village with distinct expenditure patterns. Table 12.5 compares the endowments and expenditure patterns for the two categories of HHs in the study village. Much of the intra-village variations in expenditures can be explained by this categorization. It may be mentioned that HHs participating in tourism also resort to other occupations throughout the year as tourism is an extremely seasonal activity.

The intra-village comparison confirms that HHs participating in tourism are better off in terms of per capita expenditure. They consume more than the bare necessities, as shown by the increased expenditure on non-food items and they have a positive attitude towards children's education when compared to the rest. This is in

[10]The voter-list (2005) of that part of the village lists 939 adults. No child is engaged in tourism related trade.

Table 12.5:   Household characteristics across 'tourism participants' and 'others'[†]

	HHs not directly participating in tourism	HHs directly participating in tourism
No. of HHs	38	10
Average Family size	5.3	6.0
Average age of the head of the HH	47.5	41.7*
Average Per capita landholding (katha)	7.9	5.9
% of landless or marginal HHs	66	80
% of HHs having livestock	89	90
% of literacy among adults	79	68*
% of adults completing primary education	56	31***
% of HHs directly exploiting the forest (engaged in fishing and/or prawn-fry collection)	55	30 *
% of HHs undertaking some agricultural activity	66	60
Per capita monthly expenditure on Food (Rs)	272	320
Per capita monthly expenditure on Non-food(Rs)	167	215
Per child monthly expenditure on Education[††] (Rs)	60	111*

Notes: [†] Table accounts for 48 sample households in the study village;
[††] Only HHs with children in the age group 6–18 years are considered;
*, *** indicates differences are significant at 10 per cent, 5 per cent, and 1 per cent levels respectively.

spite of their larger average family size, lower per capita landholding, and significantly lower literacy and primary education completion rate among adults. It is also found that the average age of the head of the household is significantly lower for tourism participant HHs compared to others. This may indicate that the new earning opportunities thrown up by tourism have been availed of by younger households, perhaps with a higher level of enterprise to explore the opportunities.

Table 12.5 also reveals other year-long livelihood options of tourism participant households. The percentage of HHs resorting to agriculture is not significantly lower for tourism participants. However, in spite of lower endowments in terms of physical and human capital, tourism participants show a significantly lower direct forest dependence. It may be said that tourism has provided the local 'less endowed' HHs with alternative earning opportunities which have reduced their forest dependence.

Table 12.6:   List of variables used in regressions

Variable type	Symbol used	Description
Demographic	FSIZE	Family size: number of members in the HH
	NADLT	Non-adults: Number of HH members below 18 years
	NCHLDPROP	Proportion of HH members who are not children (below 10 years)
	LARGEHH	Large HH (dummy) = 1 if HH size is greater than 5 (median size)
Physical capital	PCLAND	Per capita landholding: HH
	SOMELAND	Having some land which is not negligible (dummy) =1 if HH possesses at least 10 katha (1/15 Hectare)† of agricultural land
	LVSTOCK	Livestock (dummy) =1 if HH has livestock
Human capital	LITPROP	Proportion of literate adults to total number of adult members in the HH
Participation	TRSMDMY	Dummy = 1 if any of the HH member is engaged in tourism related job.
	FRSTDMY	Dummy 1 = if any of the HH member is engaged in direct forest exploitation.
Time and place	SEASON	Dummy, = 1 Winter (peak); 0 = Summer
	VLGDMY	Dummy = 1 if HH belongs to the study village; 0 = Control village

Note: †This figure is arrived at by discussions with local households. Landholding below this level doesn't provide any perceptible yield to the HH.

Effects of tourism on household welfare—the regression results

Though mean comparisons are helpful in bringing out differences in characteristics across social groups, the differential effect of tourism on HH welfare, both direct and indirect, can be statistically established through a set of regression analyses. Following the methodology described earlier, the variables used in regressions in this study are described in Table 12.6. The regressors in each of the estimated expenditure equations, apart from dummy variables for village and season, can be classified as variables relating to the HH's demographic composition, physical assets, human capital, and participation status with respect to tourism and forest exploitation. Participation in agriculture is not separately accommodated by a dummy variable as it is expected to be reflected by the HH's landholding status.

Table 12.7:   Regression result following OLS (robust standard errors)

(*Dependent variable: Per capita monthly monetary expenditure*[†], *PCE*)

Regressor	Coeff.	t-value
FSIZE	−31.867	−2.82***
FSIZESQR	2.325	3.77***
NCHLDPROP	105.422	2.25**
LARGEHH	−34.623	−1.74*
PCLAND	8.124	3.42***
PCLANDSQR	−0.142	−3.29***
SOMELAND	−58.977	−2.46**
LVSTOCK	−19.750	−0.80
LITPROP	28.422	0.99
TRSMDMY	50.477	1.96**
FRSTDMY	−4.566	−0.29
SEASON	−19.392	−1.47
VLGDMY	14.069	0.81
CONSTANT	269.785	3.68***
	$F_{(13, 160)}$ = 7.49	
Total observations: 174	Prob > F = 0.0000	
	R-squared = 0.2964	

Notes: [†]Including imputed values for self-produced items.

The livelihood effect of tourism is investigated first by an OLS regression of per capita monthly monetary expenditure[11] on HH level variables described in Table 12.7. It was found that 'family size' and 'per capita landholding' provide best-fit as regressors in a quadratic form. All other regressors show linear fit in the model. The dummy variables were found to be so only in the intercept form. The regression result, obtained with robust estimates of standard errors, is given in Table 12.7.

Table 12.8 further shows the regression results for expenditure on food, non-food items, and per child monthly expenditure on education. The results have been obtained by estimating the equations over the pooled data on two villages in two seasons. However, not all variables found relevance as regressors in all expenditure equations from an analytical point of view. 'Number of HH members below

[11]including imputed values for self-produced items.

Table 12.8: Regression results for expenditures on food, non-food items, and education (SURE)

Regressor	PFOOD		PNFOOD		PEE	
	Coeff.	t-value	Coeff.	t-value	Coeff.	t-value
FSIZE	–32.42	–2.42**	–35.80	–2.92***	–70.49	–4.74***
FSIZESQR	2.35	3.14***	2.85	4.17***	4.55	4.80***
NADLT	(dropped)	(dropped)	(dropped)	(dropped)	108.87	6.61***
NADLTSQR	(dropped)	(dropped)	(dropped)	(dropped)	–20.91	–5.20***
NCHLDPROP	97.29	2.07**	123.18	2.85***	(dropped)	(dropped)
LARGEHH	–34.76	–1.45	–34.13	–1.55	4.56	0.22
PCLAND	8.17	3.56***	2.36	1.12	4.85	2.49**
PCLANDSQR	–0.14	–3.08***	–0.01	–0.23	–0.10	–2.56**
SOMELAND	–59.06	–2.77***	–38.98	–2.00**	4.30	0.24
LVSTOCK	–19.90	–0.70	2.22	0.09	–24.7	–1.05
LITPROP	28.05	1.04	51.00	2.07**	–63.67	–2.84***
TRSMDMY	50.31	2.26**	60.59	2.98***	18.90	1.02
FRSTDMY	–4.39	–0.29	–2.30	–0.17	–19.23	–1.54
SEASON	–19.39	–1.53	–15.78	–1.36	2.55	0.24
VLGDMY	14.02	0.94	5.20	0.38	6.40	0.51
CONSTANT	279.14	3.81***	127.78	1.90*	220.31	5.13***
	$R^2 = 0.2963$		$R^2 = 0.3032$		$R^2 = 0.3288$	
Total observations: 174	$\chi^2 = 72.61$		$\chi^2 = 76.09$		$\chi^2 = 86.97$	
	Prob $> \chi^2 = 0.0000$		Prob $> \chi^2 = 0.0000$		Prob $> \chi^2 = 0.0000$	

Notes: Significance levels: *** at 1 per cent level; ** at 5 per cent level; * at 10 per cent level.

18 years of age' is found to be a significant determinant for 'expenditure on education' in a quadratic form. But for the other two components of expenditure, 'proportion of HH members above 10 years of age' turned out to be a more relevant regressor. Taking cognizance of the possible correlation of error terms across the components of expenditure, three equations have been estimated as a system of Seemingly Unrelated Regression Equations (SURE).

All the estimated models are found to be significant. The relatively small R^2 value can be accepted in the context of a cross-sectional HH survey data. The signs of the significant explanatory variables are mostly expected and in conformity with their economic interpretations.

Household size (FSIZE) and its square (FSIZESQR) are found to be significant for all expenditure equations with the former holding negative and the latter holding positive signs. This is expected as per capita expenditures tend to decrease for a larger HH, but the rate of decrease should slow down with additional increments in HH size.

Number of HH members below 18 years of age (the age group of students) has been included as a regressor only in explaining 'per child monthly expenditure on education'. It is also found to be significant in the quadratic form with expected sign. As the number of such members in the HH increases (perceivably from zero), per child expenditure on education increases. However, its rate of growth tends to decrease with increase in the number of such members.

The overall per capita expenditure and that for the components of food and non-food is perceived to be dependent on the proportion of HH members above 10 years of age, as they are larger consumers compared to children below that age. The higher the proportion, the greater will be such expenditures per capita. The regression results confirm this expectation.

Per capita landholding is found to be a significant regressor in quadratic form for all equations except that of 'non-food items'. This is in conformity with the heavy concentration of marginal and small farmers in the village. Land is used mostly for food (rice) for self-consumption and it provides little cash income. The self-consumption is included with its market value in expenditure on food and hence in the overall expenditure. It can be concluded that the small cash income that land may provide is significantly spent on education of children rather than on non-food items.

HHs with some viable amount of land (represented by a dummy for HHs with more than 1/15 hectare of land), are found to spend significantly lesser on per capita food and non-food items (and consequently on their overall consumption). This is apparently contradictory, but can be explained considering the feature of these villages. The landed HHs are almost entirely marginal and small. In this set-up, whenever a HH possesses some land, it tends to be tied down with the land even if the returns are very small. In contrast, HHs which are landless or have negligible landholding, are more enterprising in seeking out newer avenues of income and can make themselves better off than their poorly landed counterparts.

The proportion of literate adults in a HH is found to have an insignificant impact on per capita expenditure. However, when we consider component-wise expenditure though it is insignificant in explaining expenditure on food, it has significant positive impact on per capita expenditure on non-food items. This is in conformity with the expectation that a higher literacy rate tilts the HH's tastes in favour of non-food items. Also, its significant negative impact on per child expenditure on education may indicate that illiterates are more keen to provide education to their children. However, it is more plausible that for HHs where adults are not literates, expenditure on children tends to be more even at the primary stage as they have to be provided with local private tutors.

The dummy for forest dependents (=1 for HHs directly depending on forest-exploitation and prawn-fry collection) remains insignificant for all the estimated equations. However, the sign of its coefficient is always negative and for the education equation, it is close to being significant at the 10 per cent level. The estimated coefficients are indicative of a lower livelihood status for forest-dependents.

The dummy for tourism participation (=1 for participant HHs) is found to significantly and positively affect all regressands except education. Looking at the mean values of expenditure components in Table 12.5 and the estimated coefficient of this dummy in the corresponding equations, it can be concluded that HHs participating in tourism spend 18.5 per cent more on food and 36.3 per cent more on non-food items. As non-food items are mostly 'non-necessities' in a remote village economy, they are expected to be more income-elastic. It can be reasonably concluded that additional tourism money accruing to the participating households is spent proportionately more on non-food items after meeting their necessary food-expenditures. So, the study arrives at the conclusion that tourism significantly augments the livelihood of local participating HHs which is manifested in their expenditure pattern.

A feature of the regression results needs to be mentioned. Table 12.5 also shows that tourism participant households spend significantly more on their children's education as compared to non-participating households. However, the tourism participation dummy in the 'education' equation was not significant even at the 10 per cent level. The study recognizes this feature as a possible result of multicollinearity among the regressors. As the descriptive analysis suggests, tourism participants are poorly endowed in terms of per capita landholding and literacy, and using the participation dummy along with those regressors may have resulted in inflated standard errors of the estimated coefficients. The sign of the coefficient, however, remains expectedly positive.

The dummy indicating 'village' (=1 for study village) is not significant for any of these equations, though it has a positive coefficient for each of them. This implies that the per capita expenditures in the study village may be marginally higher but the difference is statistically not perceptible. It is indicative of the fact that the non-participating HHs (in tourism) across villages have little difference in their expenditure pattern. It can be concluded that the 'indirect' effect of tourism money in the study village, which may have accrued to the non-participating HHs through intra-village transactions, is statistically insignificant. The dummy for season (=1 for winter) is also insignificant across equations. This indicates that tourism money, earned almost entirely in the winter months by participating HHs, is retained to finance year-long expenditure and is not spent instantly in the winter months alone.

The robustness of the regression results was tested by estimating equations for HHs in the study village alone (Appendix Table A12.1.2, dropping dummy indicating village). Though the values of the estimated coefficients differ, all the regressors which were significant for the pooled data are also found to be significant for this subset of HHs, retaining similar sign. The only difference is shown by the coefficient of the dummy indicating forest-dependence in the 'education' equation. While previously it was on the threshold of significance, it in fact crosses that threshold and becomes significant at 10 per cent level (with a negative sign). This indicates an attitudinal difference for forest-dependent HHs, which spend significantly less on their children's education.

The overall impact of tourism in the study village, as brought out by the regression results, can be summarized now. HHs entering into tourism jobs in the study village are found to have significantly raised their living standard when compared to other non-participating households. The participating households are found to distribute the seasonal inflow of tourism money over their year-long expenditures. The additional money that tourism provides, enables these HHs to consume over and above the bare necessities as they are revealed to have enhanced their expenditure on non-food items proportionately more than that on food items. Trickle-down effect of tourism money to non-participating HHs by intra-village transactions is found to be statistically insignificant as such HHs show no significant increase in their expenditures by virtue of their location in the study village.

Link between tourism and conservation: empirical evidence

The regression results have established tourism participation as a significant positive determinant for improved welfare status in terms of HHs' per capita expenditures. Additionally, this study intends to look upon tourism's possible role as an alternative livelihood opportunity and hence a vehicle for conservation. In this context, a more meaningful insight can be achieved by mean comparison of HH level variables between groups classified by tourism participants and direct forest-dependents. As discussed earlier, HHs which are directly

Table 12.9:   Household characteristics across categories[12]

| | Forest-dependent households | | | | |
	Forest intruders	Prawn-fry collectors	Both forest and prawn-fry	Tourism- participants	Other Households
No. of HHs	7	19	4	10	46
Average family size	5.7	4	5.25	6	5
% of literacy among adults	56.2	78.9	81.2	67.7	82
% of adults completing primary education	37.4	51.3	50.0	30.8	66
Average per capita landholding (katha)	2.8	9.9	0.6	5.9	8.9
% of HHs having livestock	71.4	89.5	100	90	97.8
Per capita monthly consumption of food (Rs)	262	281	190	304	266
Per capita monthly consumption of non-food items (Rs)	159	177	106	211	160
Per child monthly expenditure on education[†] (Rs)	15	35	53	108	55

Note: [†]Only HHs with children in the age group 6–18 years are considered.

[12]One of the 87 HHs is found to be an outlier as a forest intruder and is dropped in this table. It has a much bigger landholding than other forest-dependents. Such cases are rare, but do occur where an erstwhile forest going HH could procure large amount of land but did not yet give up its forest going habit.

dependent on the forest for their livelihood can be categorized as: (i) solely engaged in fishing and other forest-intrusive activities; (ii) solely engaged in prawn-fry collection; and (iii) both of these activities. Some HH level characteristics across these categories, contrasted with the 'tourism participants', and 'others' are worth analysing.

Table 12.10 brings out some distinctive features among the categories based on the entire sample of the two villages. It shows that HHs resorting to both forest intrusion and prawn-fry collection are the poorest of all categories, as revealed by their average per capita monthly expenditures. This can be largely explained by their land endowments as they have the smallest per capita landholdings. In contrast, HHs which participate in prawn-fry collection only, are better-off in terms of landholding and can afford to spend as much as non-forest-dependents. However, they have a lower family size indicating lesser number of working hands. Resorting to prawn-fry collection is found to be mostly undertaken by the women of the households in the village-side rivers. They possibly cannot move out for other jobs to earn for the family.

Looking at the average per capita monthly expenditures, Table 12.10 shows that the forest intruders could largely compensate for their lesser landholding by forest exploitation. However, both in terms of literacy and completing primary education, they are backward. Additionally, their attitude towards education of their children is also dissimilar with others. This is revealed by the lowest per child expenditure on education among these households. So, the overall picture that we see is indicative of the fact that the forest intruders are landless poor people who are also somewhat divorced from the educated world.

Turning to tourism-participants, the table shows that they are the largest per capita spenders on all items. They have the largest average family size, lesser per capita landholding as compared to prawn-fry collectors and other households, lowest adult literacy except for forest-intruders and lowest proportion of adults completing primary education. In spite of these lower endowments, they can be seen as the largest spenders, especially in providing education to their children. This is indicative of the fact that mostly landless and marginal HHs with low human capital (literacy and education) have availed the new earning opportunities provided by tourism and have shown a significant attitudinal shift in building human capital by educating their children.

To corroborate the above findings, a census of tourism-related trades and participants has also been undertaken to find a more accurate picture of local participation. It is found that local people are engaged in providing a variety of services to tourists. Some of these require almost no education and capital (cooks, drinking water suppliers, boatmen). An account of all such participants in tourism shows that 78 per cent of them are engaged in trades/services with nil or a very small investment requirement. A more detailed listing of these services/trades and number of local people engaged therein is provided in Appendix Table A12.2.1.

In Table 12.5, it was seen that forest-dependence is significantly lower among tourism-participant HHs in the study village compared to the rest. The overall picture that emerges reveals that in a forest-dependent village, HHs are lowly endowed with land and human capital, which is also the case for tourism-participants. However, tourism money can lead such households out of their forest-dependence and hence can act as a vehicle for conservation. A statistically valid conclusion regarding causal relation, however, requires a probit analysis explaining forest dependence by tourism participation status. This study admits its failure to carry out such an analysis due to the problem of endogeneity which is difficult to address with a small data set.

Tourism's social effect: local perceptions and carrying capacity

The household survey questionnaire was supplemented by a set of questions for HHs in the study village. Some possible tourism-related social issues were raised and respondent's perceptions were elicited with 'Yes', 'No' and 'Indifferent' options. Analysis shows that all respondents perceived that the land prices on river-banks (vantage points for setting up a tourist lodge) have increased, but the opinion is divided on whether it is good or bad from their personal point of view. Most of the villagers feel that the spread of tourism has improved transport and telecommunication, road conditions, ferry services, and has helped in spreading Sundarbans' local cultural heritage and folk art into the outside world.

However, most of them are also of the opinion that tourism at its present scale does not significantly reduce the locals' forest dependence. They also feel that it has resulted in increased inequality in income in the village. Regarding some of tourism's possible adverse effects, which can be relevant for the policy authority, majority of the respondents answered in a positive manner. According to them, tourism did not:

1. create significant drinking water problem for locals;
2. create significant congestions in road and jetty;
3. invite petty criminals; and
4. result in significant land/air/water/noise pollution.

The unequal distribution of tourism-revenue in the study village, and uneven interactions of locals with visiting tourists are reflected by a divided opinion on a number of issues. These are whether:

1. effect of land price increase is good or bad;
2. local agricultural producers get better price for their products;
3. locals are adversely affected as buyers when tourist arrival pushes up local prices,
4. tourism provides better information about the outside world and modern health facilities;
5. tourists' behaviour results in cultural shock among locals; and
6. inequality in income earning opportunities resulting from tourism has adversely affected social fraternity and mutual faith.

It may be inferred that a mere increase in the number of tourists, without creating additional avenues for locals to participate (and helping them with finance and training) may aggravate the existing inequality (economically and perception-wise) and create some social tension in the future. A long-term policy prescription should make provision for more local poor to directly take part in tourism-related activities with active help and encouragement from the authority. It is pertinent to mention in this context that the tourism carrying capacity (TCC) in Sundarbans is not yet perceived as a constraint by the authority. This is evident from the absence of any number restrictions regarding the issue of permits to tourists.

CONCLUSION AND POLICY IMPLICATIONS

At the very outset of drawing conclusions, the study recognizes some of its data limitations. To compare livelihood status of tourism participants vis-à-vis forest dependents and others, sample HHs were selected randomly with stratification according to landholding. As village-based tourism is still in its early stage of development in the Indian Sundarbans, the number of tourism-participant HHs proved to be rather thin in the random sample. It is recognized that tourism participation may be an endogenous decision on the part of a HH, depending on its physical and human capital endowments. So, the regression results of the study may be improved upon by treating 'participation' as an endogenous variable. But a relatively small number of participating HHs in the resulting sample restricted this study from statistically exploring the possible determinants of participation.

Bearing with this limitation, the study concludes that with visitor arrivals in a remote village in the Indian Sundarbans, a significant number of landless and marginal HHs could spontaneously avail of additional earning opportunities with little or no capital investment. Poor literacy or educational status have not hindered the participation of locals in tourism jobs. The additional tourism money could significantly improve the livelihood status of the participating HHs as indicated by their per capita monetary expenditures. However, the study finds no significant indirect effect of tourism money on non-participating HHs. It also finds that since tourism is an entirely seasonal occupation at the present stage, no household survives through the year only on tourism. All the participating households are also engaged in some other year-long occupations. However, the households show a conscious spending pattern where the seasonal incremental income from tourism is used to finance their year-long expenditures evenly.

The study finds that tourism-participants have lower endowments in terms of landholding and literacy. This feature coincides with the category of 'direct forest-dependents'. Household-wise the two categories are not mutually exclusive. However, it was found that the proportion of forest-dependent HHs is significantly less among tourism-participants when compared to non-participating HHs.

The HHs participating in tourism show a positive attitude towards educating their children, probably imbibed from their constant interaction with urban visitors. Education, in turn, may reduce direct forest dependence for the local HHs due to better assimilation of information on other opportunities. Thus, tourism is expected to contribute to conservation in the long run through human capital building from within the community.

Further, the study finds that at its present scale of operation, tourism jobs are not plentiful and the percentage of people engaged is small. Forest dependence continues to be higher compared to tourism participation in terms of number of HHs involved. Lack of infrastructural facilities, most notably electricity, restricts tourist arrival to the winter months only. Also, all-inclusive package-tours originating at far-off places provide little scope for locals to enter into trade with visitors. At present, such visitors constitute the majority and hence only a smaller fraction of visitors provide a market for locals to participate in. With improvements in infrastructure, however, visitors outside such packages can grow in number and the scope for locals' participation can increase.

These conclusions lead to their attendant policy implications. While tourism shows considerable empirical evidence for livelihood improvement at the HH level, the effect at the village level is insignificant because of low participation rate of the locals. This calls for a two-pronged approach of increasing the number of visitors as well as increasing the scope for locals' participation as tourism-related service providers.

Increasing the number of visitors calls for provision of infrastructural facilities such as electricity, and improved transport and communication facilities. Also, more publicity and information dissemination may be recommended as standard tools of product promotion. New vistas may be explored in nature-based tourism, for example, inventing new products like forest walks, tree-top houses and others. These products, now absent, may be developed by private entrepreneurs once the authority comes up with a comprehensive tourism development policy.

However, increasing tourism in a quantitative way has to take into account the constraints in terms of the carrying capacity of the forest. A more immediate way of increasing tourism-jobs for local people should be the adherence to the principles of PPT. A policy of tourism promotion through EDCs, with financial and other possible supports, can make more room for the creation of such local jobs based on the same number of tourists. It calls for a greater scope for local stakeholders to participate in various components of the tourism supply chain.

REFERENCES

Alpizar R. Francisco (2002), *Essays on Environmental Policy-Making in Developing Countries: Applications to Costa Rica*, Economic Studies #117, Department of Economics, School of Economics and Commercial Law, Göteborg University.

Andersson Jessica (2004), *Welfare Environment and Tourism in Developing Countries*, Economic Studies #137, Department of Economics, School of Economics and Commercial Law, Göteborg University.

Ashley C, D. Roe, and O. Bennett (1999), *Sustainable Tourism and Poverty Elimination Study*: A report to the Department for International Development (DFID, UK), *http://www.propoortourism.org.uk/dfid_report.pdf*

Ashley, C., D. Roe, and H. Goodwin (2001), *Pro-poor tourism strategies: Making tourism work for the poor: A review of experience*, Pro-Poor Tourism Report 1, ODI, IIED, CRT. *http://www.propoortourism.org.uk/ppt_report.pdf*

Ashley, C., H. Goodwin, and D. Roe (2002), *The Tourism Industry and Poverty Reduction: A Business Primer*, Pro-poor tourism briefing No. 2. ODI, IIED, ICRT, *http://www.propoortourism.org.uk/final%20business%20brief.pdf*

Ashley, C. (2002), *Methodology for Pro-Poor Tourism Case Studie*, PPT Working Paper No. 10, Overseas Development Institute. *http://www.propoortourism.org.uk/10_methodology.pdf*

Buchsbaum B. D. (2004), *Ecotourism and Sustainable Development in Costa Rica*, (unpublished paper), *http://scholar.lib.vt.edu/theses/available/etd-05052004-171907/unrestricted/EcotourismCostRica.pdf*

Chopra, K. (2004), 'Economic Valuation of Biodivsersity: The case of Keoladeo National Park', in G. Kadekodi (ed.), *Environmental Economics in Practice: Selected Case Studies from India*, Oxford University Press, New Delhi

Chopra, K. and P. Kumar (2005), *Trade, Environment and Human Well-being*, Chapter 7 of 'Human Well-Being in the Sunderbans', Draft Report 2005.

Elliott, J. (2001), *Wildlife and Poverty Study: Phase One Report*, Livestock and Wildlife Advisory Group (LWAG), DFID (UK), *http://www.gm-unccd.org/field/Bilaterals/UK/Wildlife.pdf*

Goodwin, H. (2002), 'Local Community Involvement in Tourism around National Parks: Opportunities and Constraints', *Current Issues in Tourism*, Vol. 5, Nos 3 and 4, *http://www.multilingual-matters.net/cit/005/0338/cit0050338.pdf*

Goodwin, H., Kent, Parker, and Walpole (1997), *Tourism, Conservation and Sustainable Development, Vol. II, Keoladeo National Park, India*, Final Report to the Department for International Development (DFID, UK), *http://www.haroldgoodwin.info/resources/vol2.pdf*

Holland, J., M. Burian, and L. Dixey (2003), *Tourism in Poor Rural Areas*, PPT Working Paper No. 12. ODI, IIED, ICRT, *http://www.odi.org.uk/RPEG/PPT/WP12.pdf*

Kulkarni, A.P., V. M. Vaidya, and M. Phadanavis (2002), *Economics of Protected Area—A Case Study of Pench National Park*, Final Report

EERC Working Paper Series: WB-4, MOEF, IGIDR, World Bank, *http://coe.mse.ac.in/eercrep/fullrep/wetbio/WB_FR_AnjaliKulkarni.pdf* Pro-Poor Tourism *www.propoortourism.org.uk*

Santhakumar, V., A.K.E. Haque, and R. Bhattacharya (2005), 'An Economic Analysis of Mangroves in South Asia', in Mohsin Khan (ed.), *Economic development in South Asia*, Tata McGrawhill, New Delhi, pp. 369–437.

Saville, N.M. (2001), *Practical strategies for pro-poor tourism: case study of pro-poor tourism and SNV in Humla District, West Nepal*, PPT Working Paper No. 3. ODI, IIED, ICRT, *http://www.propoortourism.org.uk/nepal_cs.pdf*

Sills, Erin O'Donnell (1998), *Ecotourism as an Integrated Conservation and Development Strategy: Econometric Estimation of Demand by International Tourists and Impacts on Indigenous Households on Siberut Islands, Indonesia*, PhD Thesis, Department of Environment, Graduate School of Duke University.

Spenceley, A. (2001), *Development of National Responsible Tourism Guidelines and Indicators for South Africa*, Report to the Department of Environmental Affairs and Tourism, and the DFID (UK), *http://www.nri.org/NRET/guidelineslitrep.pdf*

Wunder, S. (1999), *Promoting Forest Conservation through Ecotourism Income: A case study from the Ecuadorian Amazon region*, Occasional Paper No. 21, Center for International Forestry Research (CIFOR, Indonesia) *http://www.cifor.cgiar.org/publications/pdf_files/OccPapers/OP-21.pdf*

——— (2000), Ecotourism and Economic Incentives—An Empirical Approach, *Ecological Economics*, Vol. 32, No. 3, pp. 465–79.

——— (2003), 'Native tourism, natural forests and local incomes on Ilha Grande, Brazil', in Gössling, S. (ed.), *Tourism and development in tropical islands: Political ecology perspectives*, Edward Elgar Publishing Ltd, Cheltenham, UK, pp. 148–77.

World Bank, *http://www.worldbank.org/lsms*

Appendix A12.1

Table A12.1.1: Distribution of HHs across MPCE class

(*NSS 61st Round vs Study sample*)

MPCE Class		Number of HHs	Number of HHs
Lower boundary (Rs)	Upper Boundary (Rs)	(NSS 61st Round) (2004–5)	(Study sample) (2005–6)
0	235	17	7
235	270	26	0
270	320	77	7
320	365	85	10
365	410	104	14
410	455	105	11
455	510	105	11
510	580	125	13
580	690	143	7
690	890	110	3
890	1155	49	3
1155	above	54	1
	Total	1000	87
	Median** (Rs.)	500	433

Note: ** Median is calculated as the representative value since the distribution is open-ended.

Sources: Level and Pattern of Consumer Expenditure, 2004–05; NSS 61st Round; NSSO; Ministry of Statistics and Programme Implementation, GoI. *http:// mospi.nic.in/mospi_nsso_rept_pubn.htm* (Table 1R)

A12.1.2:   Regression Results for Expenditure Equations (SURE): Study village alone

Regressor	PFOOD Coeff.	PFOOD t-value	PNFOOD Coeff.	PNFOOD t-value	PEE Coeff.	PEE t-value
FSIZE	−44.14	−2.26**	−60.29	−3.36***	−125.49	−5.20***
FSIZESQR	3.09	3.02***	4.16	4.41***	9.27	6.34***
NADLT	(dropped)	(dropped)	(dropped)	(dropped)	196.98	6.84***
NADLTSQR	(dropped)	(dropped)	(dropped)	(dropped)	−44.35	−7.04***
NCHLDPROP	214.02	2.81***	277.55	4.23***	(dropped)	(dropped)
LARGEHH	−26.86	−0.78	−1.16	−0.04	−5.52	−0.16
PCLAND	8.62	2.56***	2.01	0.65	7.93	2.50**
PCLANDSQR	−0.13	−1.94*	0.03	0.50	−0.16	−2.50**
SOMELAND	−82.33	−2.83***	−54.54	−2.03**	−20.66	−0.76
LVSTOCK	−2.19	−0.06	32.09	0.98	−15.60	−0.48
LITPROP	60.45	1.46	110.90	2.93***	−87.22	−2.44**
TRSMDMY	62.99	2.52**	78.96	3.44***	4.53	0.19
FRSTDMY	−6.21	−0.28	−16.13	−0.78	−34.09	−1.65*
SEASON	−21.57	−1.18	−17.49	−1.03	−4.19	−0.25
CONSTANT	196.90	1.63	12.91	0.12	338.47	4.70***

Total observations: 96

$R^2 = 0.3888$	$R^2 = 0.4343$	$R^2 = 0.3411$
$\chi^2 = 60.60$	$\chi^2 = 80.51$	$\chi^2 = 83.22$
Prob > χ^2 = 0.0000	Prob > χ^2 = 0.0000	Prob > χ^2 = 0.0000

Source: Authors.
Notes: See table 12.8.

Appendix A12.2
Detailed profile of tourism participants

The tourism-related trades/services adopted by local people have been exhaustively listed by this study as follows:
- Supplying drinking water in hotel/tourist lodge/in-transit boats/launches
- Cleaning of hotel rooms/bed sheets/maintenance/decoration of lodge premises
- Cooks in hotels/lodges and driver/helper/cook in tourist boats
- Paid tourist lodge managers/caretakers
- Owning tourist boats/lodges/huts/renting out own dwelling rooms to tourists
- Forest guide (regular pool and reserved pool)
- Temporary stall owners vending fruits/honey/fish and telephone booth owners
- Owning small/medium variety stores/tea stalls (stocking limited grocery items, snacks)
- Big grocery shops/stalls providing tea and breakfast
- Local ferry service/cycle-van puller
- Arranging food for on-shore tourists on contract.

Tourism being extremely seasonal (only 60 days of significant business), all the workers in the lodges are employed on a daily-wage basis. Some of the above mentioned trades/services are exclusively tourism-related and remain operative only in the winter months (peak season). Others also cater to the locals in off-season, but show significant improvement in their sale/profitability in the peak season.

Going by the electoral rolls of one poll booth in Pakhiralay in 2006, which covers the target riverside population entirely, and from which the entire local tourism-related service providers come, the number of adults is 944. Out of these, an estimated 77 persons (8.2 per cent) were found to be related to trades/services which are exclusively tourism-related. The number of persons engaged in trades not exclusively devoted to tourism, but significantly gained by it, are estimated to be 65 (6.9 per cent). Also, for 132 such persons, a classification of trades/services according to the financial investment requirements finds that 78 per cent adopted a trade/service where investment requirement is nil or less than Rs 5000 (approx US$ 110). It clearly shows that it is mostly the poorer villagers who have adopted tourism for augmenting their livelihood.

Table A 12.2.1:   Financial investments for tourism-related trades/services

Required investment (obtained by focus group discussions with traders)	Nature of service/trade	Number of engaged persons listed in study village
Nil	Supplying drinking water, cleaning of lodges, laundry services, cooking food, driving tourist boat, forest guide	52
Less than Rs 5000	Owning small stalls vending fruits/honey/fish, phone call centres, selling tea/breakfast, cycle-van pulling (on-shore conveyance)	52
Rs. 5,000–Rs 25,000	Owning small variety stores selling grocery/stationery items, snacks	5
Rs. 25,000–Rs 1,00,000	Renting out dwelling rooms to tourists and small unorganized tourist huts, owning medium size shops with grocery/stationery items, owning tourist boats	17
More than Rs 1,00,000	Owning large tourist lodge, big variety stores	6
	TOTAL	132

Source: Authors.

Contributors

CHETAN AGARWAL   Winrock International India, India

JAIME AMEZAGA   Newcastle University, UK

NESHA BEHARRY   The University of Leeds, UK

MAMTA BORGOYARY   Winrock International India, India

JEFFERY D. CONNOR   Commonwealth Scientific and Industrial Research
Organisation (CSIRO), Australia

MANUELA GAEHWILER   Eldgenossische Technische Hachshule Zürich,
Switzerland

AYELE GELAN   Consultative Group on International Agricultural Research
(CGIAR) Nairobi, Kenya

SANTADAS GHOSH   Visva-Bharati, Shantiniketan, India

CARLO GIUPPONI   Center for Environmental Economics and Management—
DSE, Venice

ALESSANDRA GORIA   Fondazione Eni Enrico Mattei, Milan, Italy

BRIAN GROSS   Department of Agricultural and Resource Economics,
Berkeley, USA

INDRILA GUHA   Vidyasagar College for Women, Kolkata, India

ROBERT A. HOPE   Oxford University Centre for the Environment, UK

FRED H. JOHNSEN   Norwegian University of Life Sciences, Norway

MUO J. KASINA   Kenya Agricultural Research Institute, Kenya

THOMAS KÖLLNER   Eldgenossische Technische Hachshule, Switzerland

PUSHPAM KUMAR   Institute for Sustainable Water, Integrated Management
& Ecosystem Research (SWIMMER), University of Liverpool, UK

ANNA LUKASIEWICZ   Commonwealth Scientific and Industrial Research
Organisation (CSIRO), Australia

DARLA HATTON MACDONALD   Commonwealth Scientific and Industrial
Research Organisation (CSIRO), Australia

ANIL MARKANDYA Fondazione Eni Enrico Mattei, Milan, Italy

JOHN MBURU Center for Development Research (ZEF), Germany

KARIN HOLM-MUELLER Center for Development Research (ZEF), Germany

ROLDAN MURADIAN Development Research Institute (IVO), The Netherlands

DIWAKAR POUDEL Local Initiatives for Biodiversity, Research and Development, Nepal

WENDY PROCTOR Land and Water Commonwealth Scientific and Industrial Research Organisation (CSIRO), Australia

RICCARDO SCARPA Waikato Management School, New Zealand

ALESSANDRA SGOBBI European Commission Europe Aid Co-operation Office, Belgium

SUNANDAN TIWARI Winrock International India, India

JOHN WARD Commonwealth Scientific and Industrial Research Organisation (CSIRO), Australia

DAVID ZILBERMAN Department of Agricultural and Resource Economics, California, USA

INDEX